P9-CAG-990

Also by Bill McKibben

The End of Nature

The Age of Missing Information

Maybe One

Long Distance: A Year of Living Strenuously

The Comforting Whirlwind

Hundred Dollar Holiday

Hope, Human and Wild

Enough

BILL McKIBBEN

Enough

Staying Human in an Engineered Age

TIMES BOOKS
HENRY HOLT AND COMPANY, NEW YORK

Henry Holt and Company, LLC
Publishers since 1866
115 West 18th Street
New York, New York 10011

Henry Holt® is a registered trademark of Henry Holt and Company, LLC.

Library of Congress Cataloging-in-Publication Data

McKibben, Bill.
Enough : staying human in an engineered age / Bill McKibben—1st ed.
p. cm.
Includes bibliographical references.
ISBN: 0-8050-7096-6
1. Human genetics—Social aspects. 2. Genetic engineering—Social aspects. I. Title.
QH438.7.M38 2003
303.48'3—dc21 2002041391

Henry Holt books are available for special promotions and premiums.
For details contact: Director, Special Markets.

First Edition 2003

Designed by Fritz Metsch

Printed in the United States of America
1 3 5 7 9 10 8 6 4 2

For Sam and Lisa Verhovek

and for Ray Karras

Contents

Introduction

As this book goes to press in January 2003, the world is still waiting to find out if the Raelian UFO cult has produced Earth's first cloned child or if that prize will go to one of the other teams of rogue scientists racing toward the goal. But the question of who will be first is, in the course of things, unimportant; the real issue is what will follow? Will this news open the gate to a "posthuman" world that the people described in this book now imagine—people who, at first glance, appear far more rational and sober than Rael's colleagues? Or will it be instead the news that rallies us to ward off a future filled with far more insidious developments than the Raelian's baby Eve?

Fifteen years ago I wrote a book called *The End of Nature.* It dealt with the way that one set of technologies—those that burn fossil fuel—were leading us into a new, dangerous, and impoverished relationship with the planet we'd been born onto. In a sense, this book follows from that one. It, too, posits that we stand at a threshold: if we aggressively pursue any or all of several new technologies now before us, we may alter our relationship not with the rest of nature but with ourselves. First human genetic engineering and then advanced forms of robotics and nanotechnology will call into question, often quite explicitly, our understanding of what it means to be a human being.

These technologies are so novel that we're loathe to tackle them, liable to hope they'll simply go away on their own. Fifteen years ago, genetic engineering on plants and animals was confined to the lab and human genetic engineering seemed a distant prospect. (In 1997, when Dolly the sheep was duplicated, some experts said humans would not be cloned for generations, if at all.) Nanotechnology and advanced robotics were mostly still on the shelves in the science fiction section. Now we know just how fast these techniques are developing—but they are still new and confusing. So first we will explore just how they work, and how exactly they might change the world.

And then we will move on to the questions of whether and how they might be controlled. One danger of any critical writing about technology is that it will be dismissed as Luddite. That's a glib charge, as silly as accusing someone of being a prohibitionist because he'd rather leave a barroom with a warm glow than a spinning head. I will be at some pains to point out the ways that some of these new technologies, if wisely limited, may help us solve certain problems we still face. But I will also raise a more fundamental set of questions: Is it possible that our technological reach is very nearly sufficient now? That our lives, at least in the West, are sufficiently comfortable? These aren't easy questions—they involve thinking hard about, among other things, poverty and illness. Answering them will be crucial to figuring out whether we actually want to rein ourselves in.

The other risk of debating technological issues is that some people assume the very debate is irrelevant—assume that we will inevitably develop each of these new technologies to the fullest despite what anyone says. Unlike global warming, however, this genie is not yet out of the bottle. Even Rael and his colleagues have not yet decided the human future—there is still room to limit and contain these technologies, if we decide to do so. This book, therefore, is first and foremost a passionate argument for having that debate. So far the discussion has been confined to a few scientists, a few philosophers, a few ideologues. It needs to spread widely, and quickly, and loudly. In that sense, the spate of clones, real or bogus, could be a gift, for it signals that the time has come to engage this question.

I think the stakes in this argument are absurdly high, nothing less than the meaning of being human. Must we forever grow in reach and power? Or can we, should we, ever say, "Enough"?

Enough

CHAPTER ONE

Too Much

For the first few miles of the marathon, I was still fresh enough to look around, to pay attention. I remember mostly the muffled thump of several thousand pairs of expensive sneakers padding the Ottawa pavement—an elemental sound, like surf, or wind. But as the race wore on, the herd stretched into a dozen flocks and then into a long string of solitary runners. Pretty soon each of us was off in a singular race, pitting one body against one will. By the halfway point, when all the adrenaline had worn off, the only sound left was my breath rattling in my chest. I was deep in my own private universe, completely absorbed in my own drama.

Now, this run was entirely inconsequential. For months I'd trained with the arbitrary goal of 3 hours and 20 minutes in my mind. Which is not a fast time; it's an hour and a quarter off the world record. But it would let a forty-one-year-old into the Boston Marathon. And given how fast I'd gone in training, I knew it lay at the outer edge of the possible. So it was a worthwhile target, a number to live with through one early-morning run after another, a number to multiply and divide against the readouts on the treadmill display when downpours kept me in the gym. It's rare enough in my life to have a goal so concrete and unambiguous.

By about, say, mile 23, two things were becoming clear. One, my training had worked: I'd reeled off one 7:30 mile after another. Two, my training wouldn't get me to the finish by itself. My legs were starting to slow and wobble, my knees and calves were hard pressed to lift and push at the same pace as an hour earlier. I could feel my goal slipping away, my pace dropping. With every hundred yards the race became less a physical test and more a mental one, game spirit trying to rally sagging flesh before sagging flesh could sap game spirit and convince it the time had come to walk. Someone stronger passed me, and I slipped onto her heels for a few hundred crucial yards, picking up the pace. The finish line swam into my squinted view, and I stagger-sprinted across. With 14 seconds to spare.

A photographer clicked a picture, as he would of everyone who finished. I was a cipher to him—a grimacing cipher, the 324th person to cross, an unimportant finisher in an unimportant time in an unimportant race. In the picture you can see the crowd at the finish, looking right past me toward the middle distance, waiting for their mom or dad, son or daughter to hove into sight. It mattered not at all what I had done.

But it mattered to me. When it was done, I had a clearer sense of myself, of my power and my frailty. For a period of hours, and especially those last gritty miles, I had been absolutely, utterly *present,* the moments desperately, magnificently clarified. As meaningless as it was to the world, that's how meaning*ful* it was to me. I met parts of myself I'd never been introduced to before, glimpsed more clearly strengths and flaws I'd half suspected. A marathon peels you down toward your core for a little while, gets past the defenses we erect even against ourselves. That's the high that draws you back for the next race, a centering elation shared by people who finished an hour ahead and two hours behind me. And it must echo in some small way what runners must always have felt—the Tarahumara

Indians on their impossible week-long runs through the canyons of Mexico, the Masai on their game trails. Few things are more basic than running.

And yet it is entirely possible that we will be among the last generations to feel that power and that frailty. Genetic science may soon offer human beings, among many other things, the power to bless their offspring with a vastly improved engine. For instance, scientists may find ways to dramatically increase the amount of oxygen that blood can carry. When that happens, we will, though not quite as Isaiah envisioned, be able to run and not grow weary.

This is one small item on the long list of "improvements" that the proponents of human genetic engineering envision, and one of the least significant corners of human life they propose to alter. But it serves as a decent template for starting to think about all the changes they have in mind, and indeed the changes that may result from a suite of other new engineering marvels like advanced robotics and nanotechnology. We will soon double back and describe the particulars of these technologies. But first consider sports.

Attempts to alter the human body are nothing new in sports, of course. It's been more than a century since Charles-Edouard Brown-Sequard, the French physiologist called "the father of steroids," injected himself with an extract derived from the testicles of a guinea pig and a dog.[1] Athletes have been irradiated and surgically implanted with monkey glands; they have weight-trained with special regimens designed to increase mitochondria in muscle cells and have lived in special trailers pressurized to simulate high altitudes.[2] For endurance athletes, the drug of choice has for the last decade been erythropoietin, or EPO, a man-made version of a hormone released by the kidneys that stimulates the production of red blood cells, so that the blood can carry extra oxygen. With EPO, the red blood cells can get so thick that the blood curdles, turns into a syrupy ooze—in the early days of the drug, elite cyclists started dropping dead across their handlebars, their hearts unable to pump the sludge running through their veins.

In 1995, researchers asked two hundred Olympic hopefuls if they'd take a drug that would guarantee them a five-year winning streak and then kill them. Almost half said yes.[3] The Tour de France has been interrupted by police raids time and again; in 2001, Italian officials found what they descibed as a "mobile hospital" trailing the Giro d'Italia bike race, well stocked with testosterone, human growth hormone, urofillitophin, salbutamol, and a synthetic blood product called HemAssist.[4] The British sports commentator Simon Eassom said recently that the only people likely to be caught for steroid abuse were from Third World countries: everyone else could afford new-generation drugs that didn't yet show up on tests.[5] Some sports, like power lifting, have had to give in and set up "drug-free" or "natural" divisions.[6]

In other words, you could almost say that it makes no difference whether athletes of the future are genetically engineered—that the damage is already done with conventional drugs, the line already crossed. You could almost say that, but not quite. Because in fact, in the last couple of years the testing has gotten better. The new World Anti-Doping Agency has caught enough offenders to throw a scare into dirty athletes, and some heart into clean ones. Some distance athletes who had decided to retire because they felt they couldn't compete have gone back into training; a new group of poststeroids shotputters and discus hurlers have proved their point by winning meets with shorter throws than the records of a decade ago.[7] And both athlete and fan remain able to draw the line in their minds: no one thought Ben Johnson's 1988 dash record meant anything once the Olympic lab found steroids in his system. It was erased from the record books, and he was banned from competition. Against the odds, sports just manages to stay "real."

But what if, instead of crudely cheating with hypodermics, we began to literally program children before they were born to become great athletes? "Picture this," writes one British journalist. "It is 2016. A young couple are sitting in a doctor's waiting room. They know that what they are about to do is illegal, but they are determined. They have come to make their child a world-beating athlete," by injecting their embryo with the patented genes of a champion.[8] Muscle size, oxygen uptake, respiration—much of an athlete's inherent capacity derives from her genes. What she makes of it depends on her heart and mind, of course, as well as on the accidents of where she's born, and what kind of diet she gets, and whether the local rulers believe that girls should be out running. And her genes aren't entirely random: perhaps her parents were attracted to each other in the first place because both were athletes, or because they were not. But all those variables fit within our idea of fate. Flipping through the clinic catalogue for athletic genes does not; it's a door into another world.

If it happens—and when that girl grows up to compete—it won't be as if she is "cheating." "What if you're born with something having been done to you?" asks the Olympic dash champion Maurice Greene. "You didn't have anything to do with it."[9] But if that happens, what will be the point of running? "Just what human excellences are we supposed to be celebrating?" asks the medical ethicist Eric Juengst. "Who's got the better biotech sponsor?"[10]

Soon, says Simon Eassom, most sport may become Evel Knievel–ish pageantry: "'Roll up, roll up, let's see somebody who'll break six seconds for the hundred meters.'" Spectacle will survive, and for many fans that may be enough. But the emptiness will be real.

To get a small sense of what it will feel like, consider the 2002 Winter Olympics, in Salt Lake City. While the North American media obsessed over figure skating disputes, the highest drama may have come on the Nordic skiing trails. Erling Jevne of Norway, a grand old man of the sport, was readying himself for one last race, the 50-kilometer, the marathon of winter. He was the sentimental favorite, in part because he had one of those sad stories that, were he an American, would have earned him hours of maudlin airtime. Raking hay on his fifth-generation family farm one day, he'd watched helplessly as his four-year-old son climbed a fence, stumbled onto a road, and was killed by a car. "I don't have a single workout where I don't think about Erich Iver," he said before the Games. "Yes, I would go far enough to say that he is an inner inspiration for my training now"—which makes Jevne not so different from all the thousands of people who run marathons in honor of their mothers, their fathers, their sons, their daughters, their friends who have died before their time or live amidst tragedy.[11] Half the people running next to me in Ottawa seemed to be wearing T-shirts with the image of some dead or dying relative.

Once before Jevne had won Olympic silver, losing to a Finn who, years later, was caught doping. This was his final stand—and he was crushed. Not long after the start, the Spaniard Johann Muehlegg caught up with him and cruised past. "His pace was simply too fast for me. He skied faster than I've ever done in my life," said Jevne.[12] As one commentator put it, Muehlegg "looked like he was skiing on another planet."[13] As indeed he was—the Planet NESP, a new EPO derivative discovered in his urine right after the race. He was stripped of his medal, although he's still appealing.

Before he heard the news—when he thought he'd simply been passed by a stronger man, or one who'd trained harder—Jevne said, "I'll recover from the disappointment. It's after all just a skiing race."[14] Which is, I suppose, the right way to think about it; for those of us who will never win a race, it should be easy to nod. But as we move into this new world of genetic engineering, we won't simply lose races, we'll lose *racing*: we'll lose the possibility of the test, the challenge, the celebration that athletics represents. Forget elite athletes—they drip one drop of sweat for every thousand that roll off the brows of weekend warriors. It's the average human, once "improved," who will have no more reason for running marathons. Say you've reached Mile 23, and you're feeling strong. Is it because of your hard training and your character, or because the gene pack inside you is pumping out more red blood cells than your body knows what to do with? Will anyone be impressed with your dedication? More to the point, will *you* be impressed with your dedication? Will you know what part of it is you, and what part is your upgrade? Right now we think of our bodies (and our minds) as givens; we think of them as us, and we work to make of them what we can. But if they become *equipment*—if your heart and lungs (and eventually your character) are a product of engineering—then running becomes like driving. Driving can be fun, and goodness knows there are people who care passionately about their cars, who will come to blows on the question Ford vs. Chevy. But the skill, the engagement, the meaning reside mostly in those who design the machines. No one goes out and drives in honor of a dying sister.

Sport is the canary in a miner's cage. It's possible the canary will die; there are those who think, with good reason, that genetic engineering of the human organism may be crude and dangerous, especially at first. But the even greater danger is that the canary will be souped up into an ever perkier, ever tougher, ever "better" specimen. Not a canary anymore, but a parrot, or a golden eagle, or some grand thing we can only guess at. A canary so big and strong that it . . . won't be a canary anymore. It will be something else entirely, unable to carry the sweet tune it grew up singing.

No one *needs* to run in the twenty-first century. Running is an outlet for spirit, for finding out who you are, no more mandatory than art or music. It is a voluntary beauty, a grace. And it turns out to be a fragile beauty. Its significance depends on the limitations and wonders of our bodies as we have known them. Why would you sign up for a marathon if it was a test of the alterations some embryologist had made in you, and in a million others? If 3 hours and 20 minutes was your design spec? We'll still be able to run hard; doubtless we'll even hurt. It's not the personal *challenge* that will disappear. It's the *personal.*

The ease with which the power of even something so peripheral can be undercut should give us pause as we move from sport closer to the center of human meaning.

In the spring of 1953, two young academics, James Watson and Francis Crick, published a one-page article in *Nature* entitled "A Structure for Deoxyribose Nucleic Acid." With it they set off the dynamite whose boom is still reverberating. In fact, the echoes grow louder; what was then theory is now becoming practice, first with plants, then with animals, and—oh so close—now with people.

"Genetics" is not some scary bogeyman. Most of the science that stems from our understanding of DNA is, simply put, marvelous—cancer drugs target tumors more effectively because now we understand much about the genetic makeup of tumor cells. But one branch of the science that flows down from the discovery of the double helix raises much harder questions. In fact, it raises the possibility that we will engineer ourselves out of existence.

We unconsciously avoid thinking about genetics too deeply, on the grounds that this science is too complicated for mere laymen, that distinctions between its various branches are surely impossible for normal folk to make. But though you may need years of study to prepare you to *conduct* genetic research, the basic science is easily understandable. And once you understand it, you have every bit as much insight into whether we should proceed with genetic alterations as any Nobel winner. The essential facts are as follows.

Genes reside in the spherical nucleus of each cell of a plant or animal; from that post they instruct the cells to make particular proteins. Those proteins, in turn, key the cell to grow or stop growing, tell it what shape to take, and so on. Grow hair. Make more dopamine. The ways in which we differ, one from another, depend in part on these blueprints, which are inside us from the moment of conception ("nature"), and in part on our experiences in the womb and after birth ("nurture").

Geneticists care about those differences that come from nature—the different pairings of DNA that cue the production of different proteins and hence different people. Some of those differences we classify as genetic disease: an errant instruction from the genes is causing a flood or drought of some protein, and hence a person develops Down syndrome, or cystic fibrosis, or any of a thousand other diseases, most of them rare and many of them devastating. Others of those differences are just *differences:* some people are taller than others, or smarter, and not just because they ate a better diet or read more books. Right up until this decade, the genes that humans carried in their bodies were exclusively the result of chance—of how the genes of the sperm and the egg, the father and the mother, combined. The only way you could intervene in the process was by choosing who you would mate with—and that was as much wishful thinking as anything else, as generation upon generation of surprised parents have discovered.

But that is changing. We now know two different methods to change human genes. The first, and less controversial, is called somatic gene therapy, and the term is one of precisely two pieces of technical vocabulary you will need to make sense of this discussion. Somatic gene therapy begins with an existing individual—someone with, say, cystic fibrosis. Researchers try to deliver new, modified genes to some of her cells, usually by putting the genes aboard viruses they inject into the patient, hoping that the viruses will infect the cells and thereby transmit the genes. If the therapy works, the proteins causing the cystic fibrosis should diminish, and with them some of the horrible symptoms. No more mucus filling the lungs, no more hopeless cough, no more drowning in your own fluid.

Somatic gene therapy is, in other words, much like medicine. You take an existing patient with an existing condition, and you in essence try to convince her cells to manufacture the medicine she needs. Such a therapy doesn't attempt to change every cell in her body, just the specific type of cells that would be transplanted. The cells of her lung tissue, say. And if she has children, the modified genes aren't passed along; when she dies, they die. Somatic gene therapy could be misused; just as athletes, for instance, misuse medicines to improve performance, so they could inject viruses with genetic materials designed to make their blood carry more oxygen or their muscles grow larger. But, as we shall see later, this is a kind of misuse we know how to deal with, or at least have a frame of reference for. No one I've ever talked to out-and-out opposes somatic gene therapy, and most wish it well. The first trials on a variety of diseases began in 1991; the first real cures were reported in 2001 and 2002; as our understanding of the human genome grows, somatic gene therapy may become more effective. It's not a silver bullet against disease, but it is a bullet nonetheless, one more item of ordnance in the medical arsenal.

"Germline" genetic engineering (and that is the other technical term), on the other hand, is something very novel indeed. "Germ" here refers not to microbes, but to the egg and sperm cells, the "germ" cells of the human being, the basic cells from which we "germinate." Scientists intent on genetic engineering would probably start with a fertilized embryo a week or so old. They would tease apart the cells of that embryo, and then, selecting one, they would add to, delete, or modify some of its genes. They could also insert artificial chromosomes containing predesigned genes. They would then take the cell, place it inside an egg whose nucleus had been removed, and implant the resulting new embryo inside a woman. The embryo would, if all went according to plan, grow into a genetically engineered child. His genes would be pushing out proteins to meet the particular choices made by his parents, and by the companies and clinicians they bought the genes from. Instead of coming solely from the combination of his parents, and thus the combination of their parents, and so on back through time, those genes could come from any other person, or any other plant or animal, or out of the thin blue sky. And once implanted they will pass to his children, and on into time. Does it sound far-fetched? We began doing it with animals (mice) in 1978, and we've managed the trick with most of the obvious mammals, except one. And the only thing holding us back is a thin tissue of ethical guidelines, which some scientists and politicians are working hard to overturn.

You could, theoretically, use this germline technique to prevent genetic disease: you could remove, from the embryonic DNA, the mistake that causes the genes to produce the cystic fibrosis proteins. But this is unnecessary. As we shall see later, if you've already isolated fertilized embryos, you can simply screen them to see which ones will naturally develop cystic fibrosis, and implant the others instead. No, the reason for performing germline genetic engineering is precisely to "improve" human beings—to modify the genes affecting everything from obesity to intelligence, eye color to gray matter. "Going for perfection," in the words of the DNA pioneer James Watson. "Who wants an ugly baby?"[15] Some of the improvements might sound "medical"—increased resistance to disease, say. But they "treat" illnesses the patient doesn't have, and, as again we shall see later, there's no way to prevent willy-nilly "enhancement" once you've started down this path. The gravitational force that we call civilization is just strong enough to hold somatic gene therapy within its orbit, but germline genetic engineering is power of another order of magnitude—a warp drive, not a nuclear reactor. It will break us free from the bounds of our past and present and send us winging off into parts unknown. That's precisely why it appeals to some.

To make germline engineering work, however, you need one more piece of technology: the ability to clone people. "Cloning" is the one part of this vocabulary that most people already know, and the one thing that scares them. In a way that's a mistake—cloning people is a sideshow, a parlor trick. Who besides rich freaks, and perhaps the grieving parents of dead children, would want exact copies?

The answer is: people who want to do germline genetic engineering. The technique of modifying genes is hard; the success rate is low. If you had more embryos, your odds would improve. That's what the people who cloned Dolly the sheep were aiming for: easy access to more embryos so they could "transform" the animals.[16] Here's how Richard Hayes, the director of the Center for Genetics and Society, and an opponent of genetic engineering, describes it: "It's very difficult to get a desired new gene into a fertilized egg on a single try. To use germline engineering as a routine procedure you'd start by creating a large culture of embryonic cells derived from a fertilized egg, douse these with viruses carrying the desired new gene," and then implant one of the eggs where the modifications worked. "Without embryo cloning, no commercial designer babies."[17] And here's a leading commentator, the Princeton biologist Lee Silver: "Without cloning, genetic engineering is simply science fiction. But with cloning, genetic engineering moves into the realm of reality."[18]

Again, it's not as if cloning is far off, or impossibly difficult. As this book went to press, the jury was still out on whether Rael and his fellow UFO cultists had actually accomplished the trick, but in any event several teams of researchers are hard at work. A few flimsy pieces of legislation are all that prevent "reproductive" cloning in most (but not all) Western nations. We've been cloning frogs for four decades; Dolly was the first mammal cloned from an adult cell, but not the last.[19] With humans "it's simply a numbers game," says George Seidel, a cloning expert at Colorado State University. "It's very likely that if you did it enough times you could make it work."[20]

But all this work will require one large change in our current way of doing business. Instead of making babies by making love, we will have to move conception to the laboratory. You need to have the embryo out where you can work on it—to make the necessary copies, try to add or delete genes, and then implant the one that seems likely to turn out best. Gregory Stock, who is a researcher at the University of California and an apostle of the new genetic technologies, says that "the union of egg and sperm from two individuals . . . would be too unpredictable with intercourse. But laboratory conception may not be a burden because such parents will probably want the most up-to-date chromosome enhancements anyway."[21] And once you've got the embryo out on the lab bench, gravity disappears altogether. "Ultimately," says Michael West, the CEO of Advanced Cell Technology, the firm furthest out on the cutting edge of these technologies, "the dream of biologists is to have the sequence of DNA, the programming code of life, and to be able to edit it the way you can a document on a word processor."[22]

All of this is new and unsettling enough that, rather than confront it head-on, people often look for a way out. A common escape hatch, especially for liberals, lies in the politically palatable notion that genes aren't all that important anyhow. We're the products of our environment, so who cares how much cutting and splicing the lab boys do?

Thankfully, there's some truth in that observation. President Bill Clinton marked the completion of the Human Genome Project sequencing by declaring, "Today we are learning the language in which God created life," but in fact creation was written in many alphabets.[23] As Francis Collins, the director of the National Human Genome Research Institute, wrote at the time, "We have seen nothing in recent studies to suggest that nature's role in development is larger, or nurture's smaller, than we previously thought."[24] Not even conditions that seem straightforwardly genetic follow some unvarying Mendelian score; sickle-cell anemia, for instance, which was formerly considered the classic single-gene disease, turns out to come in several strengths and varieties.

In some ways, in fact, the sequencing of the human genome, heralded as the dawn of the genetic age, may really have marked the sunset of a certain kind of genetic innocence. Instead of finding the expected 100,000 genes, the two teams of competing researchers managed to identify just 30,000. This total is still being debated, but whatever the final count, we have barely twice as many genes as the fruit fly, and only slightly more than the mustard weed—which makes it unlikely that genes work quite as simply as the standard models insisted.[25] Indeed, said Craig Venter, who led one of the research efforts, the small number of genes "supports the notion that we are not hard-wired. We now know the notion that one gene leads to one protein, and perhaps one disease, is false. One gene leads to many different protein products that can change dramatically once they are produced."[26] Enterprising academics in fact were quickly calling for research money to catalogue all the new proteins, an even bigger job than the genome work they'd just completed.[27] Meanwhile, those 30,000 genes, though "sequenced," were not understood. Imagine copying the works of Shakespeare by stringing all the words together without spacing or punctuation marks, said the biologist Ruth Hubbard. Then imagine handing that manuscript "to someone who doesn't know English."[28] And the traits that might interest us most—intelligence, aggression—are probably the most complicated and hidden.[29]

Plenty of practical complications make this work harder than editing text on a word processor, too. One researcher told of three hundred attempts to clone monkeys without success—"this process is just so complex," she said, with possibilities for damage right from the moment you harvest the DNA from a cell to begin work.[30] Even if you could perfect the process, simple physics would place some limits on how much you could modify humans. "If you had a nine-foot-tall person," says Stuart Newman, a researcher at New York Medical College, "the bone density would have to increase to such a degree that it might outstrain the body's capability to handle calcium."[31]

All of which sounds comforting: maybe there's not so much to worry about; maybe it's a problem for the grandkids. In fact, however, all these qualifications mask the larger truth: *genes do matter*. A lot. That fact may not fit every ideology, but it does fit the data. Endless studies of twins raised separately make very clear that virtually any trait you can think of is, to some degree, linked to our genes. Intelligence? The most recent estimates show that half or more of the variability in human intelligence comes from heredity. Even the most determined opponents of genetic engineering concede as much: David King, the British editor of *GenEthics News,* writes that "genetic determinism as an ideology is wrong and pernicious, but that doesn't mean that there aren't some completely straightforward, fairly simple, or only slightly complex genetic determinations out there."[32] Richard Hayes says, "My guess is over the next decade we'll find the full spectrum of possible relations between traits and genes: some traits will be strongly influenced by genes, others will have little relation to genes at all, others will be influenced by genes in some environments but not in others. . . . On balance, the techno-eugenic agenda would move forward," unless people stopped it.[33] Stuart Newman, a few moments after explaining why a nine-foot-tall person simply wouldn't work, leaned across his lab bench and added, "But could you engineer higher intelligence? Increased athletic ability? I have no doubt you could make such changes."[34] In other words, this new world can't be wished away.

In fact, every time you turn your back it creeps a little closer. Gallops, actually, a technology growing and spreading as fast as the

Internet grew and spread. One moment you've sort of heard of it; the next moment it's everywhere.

Consider what happened with plants. A decade ago, university research farms were growing small plots of genetically modified grain and vegetables. Sometimes activists who didn't like what they were doing would come and rip the plants up, one by one. Then, all of a sudden in the mid-1990s, before anyone had paid any real attention, farmers had planted half the corn and soybean fields in America with transgenic seed. Since 1994, farmers in this country have grown 3.5 *trillion* genetically manipulated plants.[35]

Or consider animals. Since they first cloned frogs a generation ago, researchers have learned to make copies of almost everything; it's become so standard that they now need a good gimmick to get any press attention. Texas A&M, for instance, recently called in reporters to show off the first "menagerie" of cloned animals—cows, goats, and pigs. If cloning needed a poster child, it got one in February 2001, when another team of researchers at Texas A&M unveiled Cc:, the first cat clone. The work, funded by a West Coast financier who actually set out to clone his dog, Missy, was not easy; Cc: was the only surviving animal from 87 cloned embryos.[36] On the other hand, according to a university spokesman, "she is as cute as a button."[37] (Soon cat owners everywhere were phoning their neighborhood biologists. One Colorado researcher reported a call from a woman whose cat, Stinky, had died three weeks earlier. She'd stuck his carcass in the freezer and wanted to know what to do now. "I said 'I don't think you've got any hope there,'" the scientist told her. "'Take Stinky out of the freezer and bury him.'"[38])

Under more controlled conditions, however, animal cloning is moving steadily from the lab to the factory—just as with plants, the techniques are increasingly reliable enough to let scientists scale up production. You can order cloned cattle over the Net; a high school student working at a Wisconsin firm managed to clone a cow.[39] Early in 2002, a California company debuted a chip that automates the process of nuclear transfer, the key step in cloning. Whereas now the transfer requires hours of painstaking work under a microscope, "the chip should help make cloning cheap and easy enough for companies to mass-produce identical copies."[40] A North Carolina firm has figured out a similar process for "bulk-growing" chicken embryos, which may soon allow "billions of clones to be produced each year to supply chicken farms with birds that all grow at the same rate, have the same amount of meat, and taste the same."[41]

These same technologies could be used to mass-produce human embryos; "obviously it would make everyone's life easier," said a spokesman for Advanced Cell Technology, the pioneer in human cloning research. But remember: for humans, cloning is a stepping stone. Frank Perdue might be thrilled to see billions of identical chickens, but for his own kids he'd perhaps choose a different nose. Genetic *modification* is the key, and here, too, animals are showing the way. Canadian scientists, to give just one of a thousand examples, have built what they call an Enviropig—three of them, actually, named Jacques, Gordie, and Wayne, after hockey legends. Each of the pigs' cells contains mouse and bacteria DNA, designed to cut down on the amount of phosphorus in their manure and thereby enable pork producers to raise more hogs per acre.[42] Such processes have become so standard that more and more people are getting into the act. In 1999, an artist named Eduardo Kac persuaded a laboratory to rig him up a bunny whose DNA contains genes from a phosphorescent jellyfish. If you hold Alba up to a black light, she glows green from every cell in her body; Kac needed her to "interact with him in a faux living room as a piece of performance art." Why? "It is a new era, and we need a new kind of art," Kac explained. "It makes no sense to paint as we painted in caves."[43]

The animal work seems constantly to accelerate. At the turn of the century, scientists managed to stick jellyfish genes into monkey embryos—and they tried to warn people to pay attention. "What stands out," one of the researchers told the *New York Times*, "is just how simple the method is. If it is refined to be highly effective in monkeys, it could be just a short step to using it to add genes to human embryos." Indeed, said another of the researchers, "biotechnology is forging way ahead of biology, ethics, common sense. All of us think about this all the time. All of the clinicians wonder what we are doing."[44] But no one—except the venture capitalists pouring money into biotech—paid much attention. Sure enough, a year later, in January 2001, researchers at the same lab announced that they'd managed to produce not just transgenic monkey embryos but transgenic *monkeys*. A rhesus monkey named ANDi, backward for "inserted DNA" was "playing normally" with his cagemates despite the inserted jellyfish genes that made him the planet's first transgenic primate.[45]

As for the rest of us primates? In 1963, J. B. S. Haldane, one of the last century's great biologists, said he thought it would be a millennium before the human genome could be manipulated.[46] He appears to have been off by about 960 years—but then, nearly every guess about this work has been too conservative. The Princeton biologist Lee Silver offers a short tour of the folly of underestimation: "'It is impossible to determine the sequence of the human genome,' they said in 1974. 'It is impossible to alter specific genes within the embryo,' they said in 1984. 'It is impossible to read the genetic information present in single embryonic cells,' they said in 1985. . . . All these impossibilities not only became possible but were accomplished while the early naysayers were still alive."[47] As late as November 1999, *Time* magazine was still talking about the day in 2003 when the human genome would (well ahead of the original schedule) be sequenced; as it turned out, the work was finished six months after that article appeared.[48] And it's not just the research that's accelerating, but the commercialization: in 1980 it cost a hundred dollars to sequence a single base pair of genes; the price is now counted in pennies.[49] The biotech pioneer Craig Venter said in 2002 that within five years a personalized printout of an individual's genetic code would be cheap enough for anyone to buy, so you'll probably be able to afford it late next week or so. Watch your e-mail in box for special offers.[50]

As we learn more about the human genome, we also get ever better at the mechanics of handling embryos, the technical skill required for cloning and then for germline genetic engineering.

Here's a startling statistic: some fertility clinics have gotten so handy at in vitro fertilization that "the women they treat now have a better chance of getting pregnant in one cycle than fertile women relying on plain old-fashioned sex." Some clinics in Britain and the United States boast birthrates as high as 40 percent—and of course that's for couples who usually have fertility problems to begin with. By comparison, only a quarter of fertile couples manage to get pregnant with a month of unprotected sex.[51] Which means, since the technology is not so different, that cloning a human being poses no enormous technical hurdle.

In the days after Clonaid announced the birth of baby Eve, experts were divided on whether Rael, the former French sportswriter, had actually pulled it off. Some, in fact, were convinced that the first clones had actually been born secretly before. Michael Bishop, the CEO of the animal cloning firm Infigen, described an off-the-record meeting of cloning experts at Cold Spring Harbor, the lab presided over by the DNA pioneer James Watson. "One evening after dinner some of us were talking, and there was not one of us who believed it had not already happened," he said. "It is too easy. Too bloody easy."[52] And no one discounts the possibility that either teams now at work in secret locations may succeed. Severino Antinori, for instance, an Italian scientist, has already managed to make a sixty-three-year-old woman pregnant, not to mention grow human sperm in rats, and remove unejaculated sperm from a priest and use it to fertilize the egg of a surrogate, allowing the priest to become a celibate father (or perhaps a celibate Father father).[53] Advanced Cell Technology, the Massachusetts firm that is leading the charge in the field, actually did manage to produce human embryo clones in the fall of 2001, although the clones ceased to grow at the early six-cell stage.[54]

But cloning is just the warm-up act. The main event, for people involved in this work, is germline genetic engineering: not just copying but *changing*, as we've done with plants and animals. And here, too, progress is fast, on every front. Ethical guidelines promulgated by the scientific oversight boards so far prohibit actual attempts at human germline engineering, but researchers have walked right up to the line, maybe even stuck their toes a trifle over. In the spring of 2001, for instance, a fertility clinic in New Jersey impregnated fifteen women with embryos fashioned from their own eggs, their partner's sperm, and a small portion of an egg donated by a second woman. The procedure was designed to work around defects in the would-be mother's egg—but at least two of the resulting babies carried genetic material from all three "parents."[55] This wasn't germline modification in the precise sense—a deliberate attempt to alter traits in the resulting child—but it demonstrates how close we've come with very little notice.

While waiting for the go-ahead from regulators, scientists are working with animals to develop a whole arsenal of techniques for germline manipulation. In the fall of 1998, a year after Dolly was cloned, another animal emerged that may prove more significant in the long run. Lucy, a black-brown mouse birthed in the Vancouver labs of Chromos Molecular Systems, had an extra pair of chromosomes: artificial chromosomes. She passed them on to her children, and they to theirs.[56] An artificial chromosome makes germline manipulation much, much easier; instead of having to peer through a microscope at an embryo, snipping and pasting and splicing the existing DNA in an effort to add, say, a few inches to the resulting child, a lab worker could simply insert the prepackaged chromosome. It's the difference between a scratch cake and a Duncan Hines multiplied a thousand times. "It promises to transform gene therapy from the hit-and-miss methods of today into the predictable, reliable procedure that human germline manipulation will demand," says UCLA's Gregory Stock. Indeed, Chromos already markets "Satellite DNA-Based Artificial Chromosomes," or SATACs, to people wishing to create transgenic animals, and they have a human model under development.[57] No one has put one in a human embryo yet—to do so would transgress those ethical guidelines—but they have inserted them in cultured human cells, where they replicated intact.[58]

Meanwhile, researchers in Britain and California have produced "designer sperm"; investigators at Cornell have produced an "artificial womb lining" and hope to have "complete artificial wombs" within a few years"; and so on and on and on, a cascade of developments so rapid it leaves people (and politicians) as much numbed as alarmed or excited. When the genetic revolution was in its infancy thirty years ago, scientists constantly assured the public that they would never modify the species.[59] By the mid-1980s, scientific officials were rewriting the research guidelines to permit what most regarded as the innocuous use of somatic gene therapy in humans, but they assured everyone that germline work was still off-limits.[60] But now, as the technology grows ever easier, more and more researchers want to be able to use it. "What scientists and the public need to realize is how close human germline engineering may be," writes Gregory Stock.[61] "Altering the genome is essentially the endpoint of the whole genomic revolution."[62]

Stock has done as much as any man to change the rules of the game. In the spring of 1998 he organized a one-day seminar at UCLA, "Engineering the Human Germline," which he called "the first significant public forum to focus exclusively on this difficult issue."[63] It gathered the most outspoken advocates of germline engineering, who agreed, in the words of the *Nature* correspondent who covered the conference, that "it should be implemented—regardless of concern that its use might lead to an ethical morass, and perhaps even to practices such as eugenics." In the words of John Fletcher, a University of Virginia bioethicist who attended the session, germline engineering had always before been a "Rubicon not to be crossed." But no longer; "this symposium tended to dispute that premise."[64] As a result, said other observers, "the taboo on human germline engineering [which] was absolute . . . has started to shift. Once barely considered a topic for polite conversation among even the most gung-ho of scientists, germline engineering of humans is becoming so much grist to the mill of scientists gossiping around the coffee pot."[65]

Exactly when this engineering will begin, and how fast it will spread, remain open questions. Supporters and opponents have offered dozens of different timelines. Lee Silver, for instance, the Princeton biologist whose book on the subject is called *Remaking Eden,* says the first germline therapy will be done to eliminate a few obvious diseases like cystic fibrosis (although, as we shall see, the same thing could be accomplished much more easily by simply screening embryos, not changing them). But those early interventions will cause "fears to subside," and he envisions a mother leaning back in a maternity ward in 2010, rejoicing in her new son. "I knew that Max would be a boy," she said. "And while I was at it, I made sure that Max wouldn't turn out to be fat like my brother Tom." In 2050, he sees a young mother in labor comforting herself by leafing through a photo album of what her child will look like when she's sixteen: "Five feet five inches tall with a pretty face."[66] Once germline engineering begins, it might spread very very fast; as Francis Fukuyama noted recently, cheap sonograms and easy access to abortion have rapidly reduced the numbers of girls born across Asia.[67] And Marcy Darnovsky, of the Council on Genetics and Society, points out, there's a "disturbingly comfortable fit between the techno-eugenic vision" and our consumer culture, a prevailing breeze that might easily overcome our instinctive doubts about such technologies.[68] It didn't take long for everyone to get a cell phone.

And so here's where we are: the genetic modification of humans is not only possible, it's coming fast; a mix of technical progress and shifting mood means it could easily happen within the next few years.

But we haven't done it yet. For the moment we remain, if barely, a fully human species. And so we have time yet to consider, to decide, to act. Our decision—arguably the biggest decision humans will ever make—needs to arise from clearheaded appraisal, not gentle drift. I have my set of objections to raise, and as I indicated early on they are not necessarily the objections you might think. But before we get to them, before we return to the questions of meaning that I began to consider with the homely example of sports, we need to understand more fully just what these new technologies will do. We know that genetic modification is possible in principle; now we need to know what precisely the engineers have in mind. We need to take a brief tour of the likely future.

• • •

That future would begin—indeed, it has begun—with the most obvious of human characteristics, the first thing that any parent wants to know in the delivery room: Is it a boy or a girl? In eighteenth-century France, men would tie off their left testicles to guarantee a male child; the ancient Greeks tried to work the same alchemy by lying on their right side during sex, while medieval Germans, more symbolists than physiologists, would stash a hammer beneath the bed.[69] Modern Asians have gone them one better: armed with sonograms and amniocentesis, they have conducted a virtual holocaust of girls. Of eight thousand abortions performed at one Bombay clinic after parents knew the sex of the child, 7,999 were of female fetuses.[70] But these means are fairly crude compared with the evolving technology. In the late 1990s, drawing on work done with cattle, some IVF clinics started to weigh the DNA in sperm, and then sort the sperm cells carrying X chromosomes from those carrying Ys.[71] This isn't precisely germline engineering (nothing's being altered, just chosen) and there is a theoretical medical justification (the avoidance of a few rare genetic diseases that are mostly confined to males), but most of those who paid the $2,500 price at a Virginia clinic were interested in "evening out" their families.[72] In any event, the practice may matter more as a precedent than anything else. In the words of Lee Silver, "by breaching a powerful psychological barrier, it will pave the way for true designer babies, who could really turn society upside down."[73]

And what might designer babies be designed to do? Let's begin with body and move toward mind, though the line between them is almost always fuzzy.

Some of the first germline interventions might well be semimedical, aimed at eliminating what Silver calls "predispositions" toward conditions like obesity. And, indeed, these conditions almost certainly have some genetic link; the literature is slowly filling with studies showing correlations between "the sum of extremity skinfolds" and "D7S514, an anonymous marker near the leptin gene on chromosome 7q31.3" in Mexican Americans, or between "a 5-centiMorgan region around the gene for uncoupling protein 2 on chromosome

11" and "resting energy expenditure" in French Canadian families.[74] The list of such "defects" is almost endless. One researcher said, "I did an inventory of myself and discovered that I carry eight nuisance genes. Obviously I am nearsighted—you can tell by my eyeglasses. I have dry skin. I also have a hearing defect in which I have virtually zero memory for music. . . . Wouldn't it be nice if these genes didn't have to be carried forward to my descendants."[75] As that list makes clear, the line between fixing problems and "enhancing" offspring is meaningless: almost as soon as you begin, you're worrying about conditions (like the ability to remember tunes) that would never have crossed Hippocrates' mind.

Indeed, sheer handsomeness is likely to be one of the earliest aims of genetic intervention, just as sperm banks and egg donors have long accompanied their offerings with photos. (At one clinic, a couple chose dark-haired Sperm Donor 183; when clerical error left them with the sperm of Donor 83, and one of the resulting triplets was red-haired, they sued. The wife testified at trial that she could say "with probability" that the offspring of old 183 would have been more attractive than her kids.)[76] Traits like height and muscle mass clearly have strong genetic links.

But so, perhaps, do other, subtler features. A combination of genes may "predispose" one to obesity; some of the earliest research in the field attempted to link homosexuality with a gene or several genes near the end of the long arm of the X chromosome inherited from the mother.[77] None of the evidence is ironclad yet, but there's ample data from twin and adoption studies to indicate that gayness grows from as many physical roots as psychological ones. And this offers an example of just how tricky the business of "improving" people can get. On the one hand, plenty of good liberals and gay activists, who would instinctively oppose the notion that, say, criminality had a genetic component, were secretly relieved at the idea that homosexuality could be "explained" by genes: that meant it wasn't a "moral choice," that Jerry Falwell and Pat Robertson might as well be decrying lefthandedness or red hair. On the other hand, those same good liberals might find themselves in a quandary if asked whether they wanted those predisposing genes weeded out of their offspring. Given that our society still won't even allow gay people to marry, given that society still discriminates against them in diverse ways official and unofficial—wouldn't it be kinder to the kids to reduce the chance that they would find themselves in such a position?

Once you accept the idea that our bodies are essentially plastic, and that it's okay to manipulate that plastic, then, in the words of Lee Silver, "there's nothing beyond tinkering."[78] The list expands exponentially, till there's not a feature of the human body that can't be "enhanced" in some way or another. You might, say some advocates, start by improving "visual and auditory acuity," first to eliminate nearsightedness or prevent deafness, then to "improve artistic potential."[79] But why stop there? "If something has evolved elsewhere, then it is possible for us to determine its genetic basis and transfer it into the human genome," says Silver—just as we have stuck flounder genes into strawberries to keep them from freezing, and jellyfish genes into rabbits and monkeys to make them glow in the dark. "Relatively simple attributes that fall into this category include the ability to see into the ultraviolet range or the infrared range—which would greatly enhance a person's night vision. . . . More sophisticated animal attributes include the ability to distinguish and interpret thousands of different airborne molecules present at incredibly low levels through the enhanced sense of smell available to dogs and other mammals."[80]

The list of possibilities is as long as the imagination. Some plump for eyes in the back of the head on the theory that it would "make driving safer" (and if this seems preposterous, outlandish, ridiculous, ludicrous, and impossible, reflect that way back in 1995 researchers managed to produce flies with dozens of light-sensitive eyes coating their wings and legs).[81] Others are more interested in reducing the need to sleep, in four-color vision, or in sonar.[82] The writer Lauren Slater, in a memorable *Harper's Magazine* profile, described the Dartmouth Medical School professor Joseph Rosen, a senior fellow at the C. Everett Koop Institute, who has served on advisory panels for NASA, the Navy, and the American Academy of Sciences. Rosen is also a former director of the Emerging Technology Threats bureau at the U.S. Department of Defense. He begins by asking, "Why are plastic surgeons dedicated only to restoring our current notions of the conventional, as opposed to letting people explore, if they want, what the possibilities are?" He ends, at a medical conference, by pounding the table, demanding permission to "sculpt the genotype," and announcing that, were he given permission by a medical ethics board, he would try to engineer a person to have wings.[83]

My point, for the moment, is not that such changes are evil—who hasn't dreamt the flying dream? It might be cool to see in the ultraviolet. Half the people I know obsess about getting pudgy. My point is merely that our bodies, or more precisely the bodies of our children, which have always seemed to us more or less a given, are on the verge of becoming true clay.

And not just our children's bodies, but their minds as well.

In November 2001, a team of UCLA scientists released one of those sets of results that you don't necessarily want to hear. They'd taken twenty pairs of twins, half of them identical and half fraternal, and used the newest medical scanners to examine their brains. The scanner could distinguish gray matter, the areas of the brain mainly composed of the heads of nerve cells. And what they found was that each set of identical twins had virtually the same amount of it, while the fraternal twins, who share only half of their genes, varied far more widely. Or, in the words of Dr. Paul Thompson, writing in *Nature Neuroscience,* "We found that brain structure is under significant genetic control. . . . The quantity of frontal grey matter, in particular, was most similar in individuals who were genetically alike; intriguingly, these individual differences in brain structure were tightly linked with individual differences in IQ. The resulting genetic brain maps reveal a strong relationship between genes, brain structure, and behavior, suggesting highly heritable aspects of brain structure may be fundamental in determining individual differences in cognition."[84] In other words, genes help determine how much gray matter you've got, which in turn helps determine how smart you are.

In some sense, this shouldn't have come as a great shock. A few years earlier, scientists had managed to identify the first of what may be dozens of genes that may actually cause the changes that the scanners observed. Robert Plomin of the London Institute of Psychiatry and his colleagues studied two groups of children, one with an average IQ of 103, the other with a score of 136, and found differences in a single stretch of DNA in IGFR2, "an insulin-like growth factor receptor." This one strand of code alone might account for 4 points of the IQ gap between the children, by the researchers' estimate.[85] And in 1999 Princeton scientists managed to genetically engineer a strain of mice that could more quickly "find the location of a submerged platform hidden in a pool of murky water." Not only that, the "Doogie mice" were more likely to remember a mild shock to their paws, and to recognize familiar objects more quickly—they were, in the parameters of the test, "smarter." And they were smarter because their new genes caused them to produce a protein called N-methy-D-Aspartate (NDMA) receptor, a kind of protein found in nearly identical form in the human brain.[86] That is to say, we are starting to catalogue which genes control intelligence, and starting to figure out how to manipulate them. It clearly won't be simple work; the Harvard geneticist Jonathan Beckwith, speaking last spring at the American Association for the Advancement of Science, compared the genes influencing intelligence to the many instruments in a symphony—"putting together the full orchestra is way, way off," he said.[87]

Yet news of the research still made most of us uncomfortable. In part that's because every racist and xenophobe since the dawn of time has claimed some link between ancestry and aptitude. The most recent example came in the mid-1990s, when Charles Murray and Richard Herrnstein published *The Bell Curve,* purporting to show that intelligence is heritable, and that it divides along racial lines. The second claim was dubious at best; but the first—that genes equal smarts—seemed to be an idea whose time had finally come. Various specialists marshaled data from twin and adoption studies to show that anywhere from 40 percent to 75 percent of variation in intelligence was inherited, the product of nature and not nurture. A special issue of *American Psychologist* published in the aftermath of the furor found a broad agreement among researchers that half of the variation in human intelligence appears to be related to heredity.[88]

Half is not all, of course. And IQ is not the same as ability. But IQ tracks uncomfortably closely to success—to the kinds of grades you get, and how long you stay in school, and what kind of job you hold, and how much money you make.[89] The correlation is strong enough so that you could argue it might make sense to soup up your child, for either her sake or the planet's. The idea's in the air: "As society gets more complex perhaps it must select for individuals more capable of coping with its complex problems," says Daniel Koshland, a man who for ten years edited *Science,* the country's most prestigious scientific journal, and who has a hall named after him on the Berkeley campus. "If a child destined to have a permanently low IQ could be cured by replacing a gene, would anyone really argue with that? . . . It is a short step from that decision to improving a normal IQ. Is there an argument against making superior individuals?"[90] There is, I think, and I will get to it shortly—but it is an argument that will be made against the odds. Already the engineers are promising quick results: in the fall of 2000, Robert Lanza, the vice president of Advanced Cell Technology, the Massachusetts firm that was the first to clone a human embryo, told a New York forum that scientists are "close to being able to add 20 or 30 points to your baby's IQ."[91] Already the political scientists are imagining the consequences: at a 1999 symposium at the Jackson Laboratory in Bar Harbor, Maine, LeRoy Walters, director of the Kennedy Institute of Ethics at Georgetown University, urged geneticists to start thinking about establishing a baseline "for the portion of intelligence that is genetically determined," so that the government could guarantee access to genetic improvement for anyone falling below the line, lest they be doomed to join a new genetic underclass.[92]

Just as our list of potential modifications of the body began with the relatively obvious and spiraled off toward the fantastic, so with the mind. It was only a generation ago that E. O. Wilson set off enormous controversy with his book *Sociobiology*, arguing that evolution, as encoded in our genes, had helped produce our attitudes and cultures. By now, as good members of the Prozac generation, we're pretty comfortable with the notion that mood is a function of chemistry, and hence in some measure of the genes that control that chemistry. Researchers at the National Institutes of Health, for instance, have found a stretch on Chromosome 17 that predisposes people to anxiety, apparently by influencing the production of a protein that regulates the level of serotonin in the brain. People with the "sluggish" version of the gene scored higher when tested for worrying, pessimism, and fear. The gene only accounted for a small percentage of the variation, but there are almost certainly other such stretches of DNA elsewhere in our cells.[93]

As with anxiety, so with aggression. Though it's another topic that strains the bounds of political correctness, at least some violent behavior seems genetically linked. As Francis Fukuyama pointed out in his book *Our Posthuman Future*, as early as the 1980s Dutch researchers studying a family with "a history of violent disorders traced the cause to genes that control the production of enzymes known as monoamine oxidases," or MAOs. A French team studying mice showed that a similar defect in their MAO genes led them to turn "extremely violent."[94] Already some ethicists have called for judges to check particular stretches of DNA for a "predisposition" toward antisocial behavior before passing sentences.[95]

Every aspect of personality may in some way reflect your collection of proteins. Several years ago, for instance, researchers at the National Human Genome Research Institute developed a so-called knockout mouse—in this case, it was missing a gene labeled "disheveled 1." Mice lacking the gene "look normal, and achieve normal grades on tests of learning and memory," but when you put them in a cage with others mice they interact less often than normal mice, they fail to "fluff up suitable beds from their nesting material," and they can't manage to trim one another's whiskers properly. "The inattention to barbering is a sign that they do not form the social hierarchy customary among this strain of laboratory mice," reported Dr. Anthony Wynshaw-Boris. "I think this is the first gene to be described that controls social interactions."[96] Other investigators took still another gene, one from a prairie vole, and stuck it in a lab mouse—this "slightly different version of a receptor for a brain peptide associated with social behavior" produced markedly different behavior in the mouse, who suddenly was grooming and licking like . . . a vole.[97]

Such data may help clinicians trying to treat schizophrenics—indeed, last summer Irish researchers identified two genes connected with human schizophrenia—but, as usual, there's no obvious line between repair and improvement.[98] Other researchers, for instance, are already hot on the trail of a human "happiness gene," and at the moment they're concentrating on "the gene for the dopamine D4 receptor, which contains a hypervariable coding in its third exon." An Israeli group found that certain variations of the gene made people more likely to seek out novelty—and more likely to answer yes to statements such as "Sometimes I bubble with happiness" and "I am a cheerful optimist." Such hardwiring may "determine our average set-point" for happiness, the researchers argue, so that even "winning the Nobel Prize or marrying our childhood sweetheart may not alter our overall happiness—for that, gene therapy would be required."[99]

In short, it's not particularly far out to imagine genetic engineering designed to make our children happier—a kind of targeted, permanent Prozac. Dean Hamer, who is no less than the chief of gene structure and regulation at the National Cancer Institute's Laboratory of Biochemistry, imagines a future scenario in which a young couple, Syd and Kayla, get to tweak the emotional makeup of their fetus. "They pondered the choices before them, which ranged from the altruism level of Mother Teresa to the most cutthroat CEO. Typically, Syd was leaning toward sainthood; Kayla argued for an entrepreneur. In the end, they chose a level midway between, hoping for the perfect mix of benevolence and competitive edge. . . . Syd and Kayla, however, did not want to set their child's happiness rheostat too high. They wanted her to be able to feel real emotions. If there was a death, they wanted her to mourn the loss. If there was a birth, she should rejoice."[100] Or, to put it another way, here is Gregory Pence, a University of Alabama professor, in his book *Who's Afraid of Human Cloning?*: "Many people love their retrievers and their sunny dispositions around children and adults. Could people be chosen in the same way? Would it be so terrible to allow parents to at least aim for a certain type, in the same way that great breeders . . . try to match a breed of dog to the needs of a family?"[101]

By now, the vision of the would-be genetic engineers should be fairly clear. It is to do to humans what we have already done to salmon and wheat, pine trees and tomatoes. That is, to make them *better* in some way; to delete, modify, or add genes in the developing embryos so that the cells of the resulting person will produce proteins that make them taller and more muscular, or smarter and less aggressive, maybe handsome and possibly straight, perhaps sweet. Even happy. It is, in certain ways, a deeply attractive picture.

Before we decide whether all that adds up to a good idea, there's just one more factual question to be answered: Would we actually do this? We've heard from the salesmen making the case, but would we actually buy? Is there any real need to raise these questions as more than curiosities, or will the schemes simply fade away on their own, ignored by the parents who are their necessary consumers and then forgotten by history?

I grew up in a household where we were very suspicious of dented cans. Dented cans were, according to my mother, a well-established gateway to botulism, and botulism was a bad thing, worse than swimming immediately after lunch. It was one of those bad things measured in extinctions, as in "three tablespoons of botulism toxin could theoretically kill every human on earth." Or something like that.

So I refused to believe the early reports, a few years back, that socialites had begun injecting dilute strains of the toxin into their brows in an effort to temporarily remove the vertical furrow that appears between one's eyes as one ages. It sounded like a Monty Python routine, some clinic where they daubed your soles with plague germs to combat athlete's foot. But I was wrong to doubt. As the world now knows, Botox has become, in a few short years, a staple weapon in the cosmetic arsenal—so prevalent that, in the words of one writer, "it is now rare in certain social enclaves to see a woman over the age of 35 with the ability to look angry." With their facial muscles essentially paralyzed, actresses are having trouble acting; since the treatment requires periodic booster shots, doctors warn that "you could marry a woman [or a man] with a flawlessly even face and wind up with someone who four months later looks like a Shar-Pei."[102] But never mind—now you can get Botoxed in strip mall storefronts and at cocktail parties. "After a brief discussion of benefits and potential risks, everyone starts drinking," explained one doctor who hosts such soirees. "It really takes the edge off."[103]

People, in other words, will do fairly far-out things for less than pressing purposes. And more so all the time: public approval of "aesthetic surgery" has grown 50 percent in the United States in the last decade, and there's no automatic reason to think that consumers would balk because it was "genes" involved instead of, say, "toxins."[104] Especially since germline engineering would not promote your own vanity, but instead be sold as a boon to your child. Anyone who has entered a baby supply store in the last few years knows that even the soberest parents can be counted on to spend virtually unlimited sums in pursuit of successful offspring. What if the "Baby Einstein" video series, which immerses "learning enabled" babies in English, Spanish, Japanese, Hebrew, German, Russian, and French, could be bolstered with a little gene-tweaking to improve memory? What if the WombSongs prenatal music system, piping in Brahms to your waiting fetus, could be supplemented with an auditory upgrade? According to the *Wall Street Journal*, upscale parents are increasingly buying $18 bottles of baby shampoo, and massaging their infants with "Bonding Oil," an unguent which allows the youngster "to rejuvenate for another day of exploration and growth."[105] One sociologist told the *New York Times* we'd crossed the line from parenting to "product development," and even if that remark is truer in Manhattan than elsewhere, it's not hard to imagine what such attitudes will mean across the affluent world.[106] As early as 1993, a March of Dimes poll found that 43 percent of Americans would engage in genetic engineering "simply to enhance their children's looks or intelligence."[107]

Here's one small example. In the 1980s, two drug companies were awarded patents to market human growth hormone to the few thousand American children suffering from dwarfism. The FDA expected the market to be very small, so HGH was given "orphan drug status," a series of special market advantages designed to reward the manufacturers for taking on such an unattractive business. But within a few years, HGH had become one of the largest-selling drugs in the country, with half a billion dollars in sales. This was not because there'd been a sharp increase in the number of dwarfs, but because there'd been a sharp increase in the number of parents who wanted to make their slightly short children taller.[108] Before long the drug companies were arguing that the children in the bottom 5 percent of their normal height range were in fact in need of three to five shots a week of HGH.[109] Take eleven-year-old Marco Oriti. At four foot one, he was about four inches shorter than average, and projected to eventually top out at five feet four. This was enough to convince his parents to start him on a six-day-a-week HGH regimen, which will cost them $150,000 over the next four years. "You want to give your child the edge no matter what," said his mother.[110]

A few of the would-be parents out on the current cutting edge of the reproduction revolution—those who need to obtain sperm or eggs for in vitro fertilization—exhibit similar zeal. Ads started appearing in Ivy League college newspapers a few years ago: couples were willing to pay $50,000 for an egg, provided the donor was at least five feet, ten inches tall, white, and had scored 1400 or better on her SATs.[111] (A few months later, a fashion photographer opened a Web site to auction eggs from top models. He offered no guarantees concerning their board scores, saying only, "This is Darwin's natural selection at its very best—the highest bidder gets youth and beauty.")[112] There is, in other words, a market just waiting for the first clinic with a catalogue of germline modifications, a market that two California artists proved when they opened a small boutique, Gene Genies Worldwide, in a trendy part of Pasadena. Tran T. Kim-Trang and Karl Mihail wanted to get people thinking more deeply about these emerging technologies, so they outfitted their store with petri dishes and models of the double helix, and printed up brochures highlighting traits with genetic links: creativity, extroversion, thrill-seeking, criminality. When they opened the doors, they found people ready to shell out for designer families (one man insisted he wanted the survival ability of a cockroach).[113] The "store" was meant to be ironic, but the irony was lost on a culture so deeply consumer that this kind of manipulation seems like the obvious next step. "Generally, people refused to believe this store was an art project," says Tran.[114] And why not? The next store in the mall could easily have been a piercing parlor or a Botox salon. We're ready. And no one's even begun to advertise yet.

But say *you're* not ready. Say you're perfectly happy with the prospect of a child who shares the unmodified genes of you and your partner. Say you think that manipulating the DNA of your child might be dangerous, or presumptuous, or icky? How long will you be able to hold that line if the procedure begins to spread among your neighbors? Maybe not so long as you think: if germline manipulation actually does begin, it seems likely to set off a kind of biological arms race. "Suppose parents could add 30 points to their child's IQ," asks the economist Lester Thurow, of the Massachusetts Institute of Technology. "Wouldn't you want to do it? And if you don't, your child will be the stupidest in the neighborhood."[115] That's precisely what it might feel like to be the parent facing the choice. Individual competition more or less defines the society we've built, and in that context love can almost be defined as giving your kids what they need to make their way in the world. Deciding not to soup them up . . . well, it could come to seem like child abuse.

Of course, the problem with arms races is that you never really get anywhere. If everyone's adding 30 IQ points, then having an IQ of 150 won't get you any closer to Stanford than you were at the outset. The very first athlete engineered to use twice as much oxygen as the next guy will be unbeatable in the Tour de France—but in no time he'll merely be the new standard. You'll have to do what he did to be in the race, but your upgrades won't put you ahead, merely back on a level playing field. You might be able to argue that society as a whole was helped, because there was more total brainpower at work, but your kid won't be any closer to the top of the pack. All you'll be able to do is up the odds that she won't be left hopelessly far behind.

In fact, the arms-race problem has an extra ironic twist when it comes to genetic manipulation. The United States and the Soviet Union could, and did, keep adding new weapons to their arsenals over the decades. But with germline manipulation, you get only one shot: the extra chromosome you stick in your kid when he's born is the one he carries throughout his life. So let's say baby Sophie has a state-of-the-art gene job: her parents paid for the proteins discovered by, say, 2005 that, on average, yielded 10 extra IQ points. By the time Sophie is five, though, scientists will doubtless have discovered ten more genes linked to intelligence. Now anyone with a platinum card can get 20 IQ points, not to mention a memory boost and a permanent wrinkle-free brow. So by the time Sophie is twenty-five and in the job market, she's already more or less obsolete—the kids coming out of college just plain have better hardware. "For all his billions, Bill Gates could not have purchased a single genetic enhancement for his son Rory John," writes Gregory Stock. "And you can bet that any enhancements a billion dollars can buy Rory's child in 2030 will seem crude alongside those available for modest sums in 2060."[116] It's not, he adds, "so different from upgraded software. You'll want the new release."[117] The vision of one's child as a nearly useless copy of Windows 95 should make parents fight like hell to make sure we never get started down this path. But the vision gets lost easily in the gushing excitement about "improving" the opportunities for our kids.

If germline genetic engineering ever starts, it will accelerate endlessly and unstoppably into the future, as individuals make the calculation that they have no choice but to equip their kids for the world that's being made. Once the game is under way, in other words, there won't be moral decisions, only strategic ones. If the technology is going to be stopped, it will have to happen now, before it's quite begun. The choice will have to be a political one, that is—a choice we make not as parents but as citizens, not as individuals but as a whole, thinking not only about our own offspring but about everyone. And given the seductions that we've seen—the intuitively and culturally delicious prospect of a *better* child—the arguments against must be not only powerful but also deep. They'll need to resonate on the same intuitive and cultural level. We'll need to feel in our gut the reasons why, this time, we should tell Prometheus thanks, but no thanks.

I was a young newspaper reporter in Cambridge, Massachusetts, in the late 1970s and early 1980s, the years when whole labs from the Harvard and MIT biology departments were spinning off into the early biotech companies, and Nobel laureates were turning into CEOs. These were the days before every city in America was vying for its share of the biotech industry; in fact, it was not at all clear to residents and local officials that they really wanted this kind of work going on within the crowded city limits. Al Vellucci, a neighborhood politician of considerable charm who had been on the city council throughout living memory, made it his particular mission to grill the various eggheads who would assemble at one hearing after another as the city came up with its own regulations to govern genetic research. With the false smiles of men forced to seek favors from their intellectual inferiors, the scientists would explain with exaggerated patience the safety of their work, the unlikelihood of "germs" escaping through the sewers. With eyebrows lifted, they would once again illuminate for the good councilor the many reasons why Cantabrigians shouldn't fear a Frankenstein monster. And eventually they would get their way, albeit only after having made expensive and extensive concessions.

Those city hall debates come back to me as I think about one of the lines of argument against genetic research, the idea that something might go very, very wrong. Both sides in those early controversies were correct: the dangers were not as direct as the critics imagined, but recombinant DNA did pose many potential perils. Recent Australian research on mousepox inadvertently pointed the way toward a super-virulent smallpox, for instance, while experiments on the potentially disease-causing bacterium *E. coli* produced a strain 32,000 times stronger than usual.[118] Especially in the wake of September 11, 2001, people are thinking more deeply about whether it makes sense to research genes that could, say, reduce the protein content of an enemy's grain crops; indeed, shortly after the attacks, the Biotechnology Industry Organization, "acting at the government's request, asked all its member companies what technology they had that could be used to create bioweapons and asked them to be on the alert for unusual orders for their products."[119] But these kinds of fears do not indict human germline engineering in particular: there is no special reason to think that it will unleash some new plague.

Some critics have pointed out more subtle ways in which germline engineering could do damage—for instance, by altering the gene pool that all human beings share. The Parliamentary Assembly of the Council of Europe resolved that individuals have a right to a genetic heritage that has not been tampered with.[120] In fact, however, we "alter" the gene pool every time we save someone who has a disease that would otherwise kill him before he reproduced.[121] And in a world where more than 10 million babies are born each month, it will be some time before engineering affects the human genetic heritage in more than a trivial way.[122]

So the real debate centers on a very different question: what happens to the world if everything goes right—if it all works out the way the engineers imagine, and we find ourselves able to produce designer babies who grow up to be all the things their parents chose from the catalogue?

For many, the first specter that flashes to mind is eugenics. The idea of "genetic improvement," after all, reached its twentieth-century apogee in the Nazi era. Hitler's *Lebensborn* program mated "racially pure" German women with SS officers to perpetuate "desirable genotypes"; meanwhile, the authorities were busy sterilizing undesirables and, of course, trying to exterminate some races altogether.[123] There are those, particularly on the left, who fear something similar from the new genetic engineers. "Geneism" could "eclipse racism as the most destructive force on the planet," George Annas, a professor of health law at Boston University, told the World Conference Against Racism in 2001.[124] Once some of us are enhanced, he asks, "won't we see other people as subhuman, and enslave or slaughter them?"[125] But it's harder to get worked up over the idea of eugenics in an age of more limited government. Is it still "eugenics" if no one's forcing you to do it? If you're making all the choices for your own kid? Fashion may be looking over your shoulder—young women of childbearing age around the world are watching the same Pink video—but fashion isn't Hitler.

So these would be mere consumer decisions—but that also means that they would benefit the rich far more than the poor. That they would take the gap in power, wealth, and education that currently divides both our society and the world at large, and write that division into our very biology. A sixth of the American population lacks health insurance of any kind—they can't afford to go to the doctor for a *checkup*. And much of the rest of the world is far worse off. If we can't afford the fifty cents a person it would take to buy bed nets to protect most of Africa from malaria, it is unlikely we will extend to anyone but the top tax bracket these latest forms of genetic technology.

This injustice is so obvious that even the strongest proponents of genetic engineering make little attempt to deny it. The most revealing account of the divided future comes from Princeton geneticist Lee Silver, in his book *Remaking Eden.* "Emotional stability, long-term happiness, inborn talents, increased creativity, healthy bodies—these could be the starting points chosen for the children of the rich," he writes. "Obesity, heart disease, hypertension, alcoholism, mental illness—these will be the diseases left to drift randomly among the families of the underclass." He forecasts a future divided between "the GenRich" and "the Naturals." The former, perhaps 10 percent of the American population, "all carry synthetic genes," and they control "all aspects of the economy, the media, the entertainment industry, and the knowledge industry." The latter work as "low-paid service providers and laborers."[126] In this world there are

> many types of GenRich families—for example, there are GenRich athletes who can trace their descent back to professional sports players from the twenty-first century. One subtype of GenRich athlete is the GenRich football player, and a sub-subtype is the GenRich running back. Embryo selection techniques have been used to make sure that a GenRich running back has received all of the natural genes that made his unenhanced foundation ancestor excel at the position. But in addition, at each generation beyond the foundation ancestor, sophisticated genetic enhancements have accumulated so that the modern-day GenRich running back can perform in a way not conceivable for any unenhanced Natural.

There are, as well, GenRich scientists, whose set of "particular synthetic genes" produce an "enhanced scientific mind," and GenRich businessmen, and so on and so forth. "There is still some intermarriage as well as sexual intermingling between a few GenRich individuals and Naturals," Silver imagines, but "GenRich parents put intense pressure on their children not to dilute their expensive genetic endowment in this way."[127] And indeed, eventually, they become "entirely separate species, with no ability to cross-breed, and with as much romantic interest in each other as a current human would have for a chimpanzee."[128]

Silver is not alone in his vision. Joseph Fletcher, a University of Virginia philosopher sometimes called "America's patriarch of medical ethics," imagined that some kinds of hybrid ape-humans might be concocted, "chimeras or parahumans to do dangerous or demeaning jobs."[129] Princeton physicist Freeman Dyson, upon winning the Templeton Prize for Progress Toward Research or Discoveries About Spiritual Realities, warned a crowd at the National Cathedral in Washington that germline engineering "could cause a splitting of humanity into hereditary castes."[130] Anyone thinking this might not be *so* bad should recall Jim Crow America, or go to the Hindu world and see those places where the caste system still persists. I remember visiting Kerala, in the south of India, where only a century ago caste was so entrenched that a Brahmin walking on the road was preceded by a member of a slightly lower caste giving a warning shout so the Brahmin's inferiors could get out of sight. One local poet, an untouchable, described his kin this way: "They walk so gently, with fear in mind, that even the earth does not feel their tread . . . even grass would not make way before them."[131] The greatest accomplishment of our recent history, the triumph of the Gandhis and the Kings and all who worked with them, was the gradual elimination of precisely this kind of sadness.

But what if their triumph was only temporary? Francis Fukuyama, the political scientist, has thought more carefully than anyone else about the implications of these new divisions. His measured, unhysterical book *Our Posthuman Future* conjures up a truly frightening vision of a world where the successful young people owe their prospects not "to the accidents of birth and upbringing" but instead to the "good choices and planning of their parents." Success for them is "deserved," that is, not something to be grateful for. "They will look, think, act, and perhaps even feel differently from those who were not similarly chosen, and may come in time to think of themselves as different kinds of creatures. They may, in short, feel themselves to be aristocrats, and unlike aristocrats of old, their claim to better birth will be rooted in nature and not convention."[132] That would be a very different place from the America of the Founding Fathers, as Fukuyama points out—*they* believed that all men were created equal. They defined "citizen" too narrowly, of course, but in their spirit we have steadily expanded its meaning to include both genders and all races. Now, however, the definition might contract again, and with "malign consequences for liberal democracy and the nature of politics itself."[133] The political equality enshrined in the Declaration can't withstand the destruction of the idea that humans are in fact equal; what, Fukuyama asks, "will happen to political rights once we are able to, in effect, breed some people with saddles on their backs, and others with boots and spurs[?]"[134]

To this there is no good answer, only the gibe that it won't make any difference. "Anyone who accepts the right of affluent parents to provide their children with an expensive private school education cannot use 'unfairness' as a reason for rejecting the use of reprogenetic technologies," says Silver. "Whether we like it or not, the global marketplace will reign supreme."[135] Given the experience of the last few decades—a steadily widening gap between rich and poor, in this country and around the world—it's entirely possible (and entirely discouraging) that he's correct.

So far, all the arguments mustered against germline engineering have concerned its effect on society—the danger posed to our health by runaway germs, and to our liberty by runaway inequality. But there's another kind of argument as well, in some ways closer to the intuitive and visceral reasons that I'm seeking, an argument that begins with individual children whose genes would be modified.

Let's start with a pig, Pig #6707, reared at the USDA research center in Beltsville, Maryland. Scientists hoping to duplicate earlier work on mice inserted a human growth gene into the porker while it was yet an embryo; he would, they hoped, turn into a veritable mountain of bacon. And indeed, when Pig #6707 was born his pituitary gland did produce human growth hormone. Unfortunately, instead of simply causing him to get larger, "the human genetic material altered the pig's metabolism in an unpredictable and unfortunate way. Excessively hairy, lethargic, riddled with arthritis, apparently impotent, and slightly cross-eyed, the pig could hardly stand up."[136] (He did, however, have less body fat, allowing the USDA to claim it had produced "leaner meat."[137]) Meanwhile, cloned mice in a number of studies have shown a marked tendency to obesity, while other cloned mice seemed to die well ahead of schedule.[138] (Another study seemed to show that for five generations cloned mice lived out normal life spans, "but the sixth generation animal was eaten by a foster mother, bringing the experiment to an abrupt end."[139]) So the question naturally arises: if you start genetically engineering children, might you not get some excessively lethargic, obese, and hairy people too?

When Stuart Newman, the New York Medical College professor, gives a talk about cloning, he always begins with a slide of some of those gargantuan cloned mice, looming over their cagemates like Shaq visiting a kindergarten. "Large offspring syndrome happens fairly often," he says. "When scientists perform genetic modifications on a mouse, it doesn't always work out the way they expect," he notes—with the authority of someone who's spent his life surrounded by pipettes, centrifuges, incubators. "One gene might be thought to influence the size of the mouse; another expected to influence its speed in going through a maze. . . . But very often the outcomes are completely surprising," varying widely from species to species, within different strains of the same species, and even from individual to individual within a single strain.[140] "All scientists know that when you do a biological experiment it sometimes works and sometimes doesn't. If it works 70–80 percent of the time you are really lucky." And even if it appears to have worked, you won't know for generations: a germline-modified animal may appear completely unremarkable, but its progeny may, for instance, develop cancer at forty times the normal rate.[141] Given our current attitudes toward lab mice and field animals, that doesn't represent an enormous problem. You just keep experimenting until you've figured it all out, and if you have a bunch of goofy batches—Doogie mice that turn out to be Sumo mice—you throw them out and start over. But people are more problematic: if you go into the clinic for a Doogie kid and the doctor gives you a hairy, cross-eyed fatso, you're going to sue. More than that, it's clearly unethical. How could you justify the unhappy life you'd created? And indeed, no one really disagrees. Even the most enthusiastic advocates of germline manipulation say that for the moment it's still too risky; the National Bioethics Advisory Commission ruled in 1997 that such work "is not safe to use in humans at this time."[142]

But the key words in that ruling were "at this time." Newman and some others hold that such modifications will forever be unsafe and hence unethical: "How will we get the knowledge?" he asks. "Not by studying mice—the tolerances in mice are different than the tolerances in rats, not to mention people. The only way we could ever do this with people is through experimentation—only by putting in the genes and seeing what happens. And that's impermissible. No matter what happens on the other side, there's an ethically impossible passage through which you must go—the generations of humans that would be experiments."[143] But that's a passage that zeal or profit might well force open. "The mere fact that there may be unanticipated or long-term side effects will not deter people from pursuing genetic remedies, any more than it has in earlier phases of medical development," predicts Francis Fukuyama; indeed, he says, emerging computer modeling techniques may quickly make it easier to predict what will happen when you mess around with genes.[144] And many proponents of germline engineering grow increasingly impatient with the risk argument. "Nature creates thousands of monsters every day," Rael, the UFO cult leader turned would-be cloner, said last year. Why should science be held to a higher standard?[145] If Rael seems a little far out, then consider Daniel Koshland, the longtime editor of *Science,* speaking at a UCLA symposium in 1998: "There is no such thing as absolute safety in this world. . . . Most of us know that risks must be relative, and that they will be taken if the gains seem to be in some proportion to the risks. If you think it over, the whole process of conception and birth really is a very risky and dangerous proposition. My guess is that germline engineering will compete very well with those conceived the natural way."[146] In other words, don't count on the inherent riskiness of this work—the fact that it will necessarily involve experimentation on people—to bring it to a halt.

For many conservatives, the most convincing argument is perhaps a little different; they worry less that these technologies imperil, and more that they degrade. That they represent, in the words of the columnist Charles Krauthammer, "the ultimate in desensitization," "an assault on human dignity."[147] Leon Kass, the University of Chicago ethicist appointed by George W. Bush to lead the President's Council on Bioethics, says they put scientists and doctors into the "godlike role of creator, judge, and savior."[148]

These arguments draw on a powerful reflex, what Kass and James Q. Wilson have called our "instinctive recoil" at the prospect of a genetically engineered future.[149] But, at least in the form they've so far been presented in, I doubt they are strong enough. They appeal to the religious right, but they scare the majority of Americans who, for instance, support a woman's right to abortion, a process that also involves, for better and for worse, a certain amount of playing God. It's too easy to imagine that such talk is the chatter of people who don't want evolution taught in the schools, easy to say that your religious belief should not block my reproductive liberty. And in such a stalemate, the tie will probably go to the scientist, offering us "progress" and sneering at "superstition." A survey of Britain's leading researchers, for instance, found them convinced that human cloning and genetic manipulation would soon be under way. As Richard Dawkins, professor of the public understanding of science at Oxford University, put it, "People who object to research of this kind must explain exactly who would, in their view, be damaged by it. Phrases like 'playing God' form no part of a valid argument."[150]

And in any event none of these arguments quite captures the truly horrifying aspects of this new technology, the features that distinguish it from what's come before. So what follows is one more line of contention, which is neither entirely utilitarian, nor religious in the usual sense. It is, as I began to sketch in my description of the marathon, an argument from meaning.

From a certain vantage point, meaning has been in decline for a very long time, almost since the start of civilization. Our hunter-gatherer ancestors inhabited a very different world from ours, a meaning-saturated world where every plant and animal was an actor the way people are actors, where even rocks and mountains and canyons and rivers could speak. We look at that same world and see either silent landscape or pile of resources; either it has gone mute, or our hearing is nowhere near as sharp.[151]

But the context of our lives began to shrink much more quickly in the last five hundred years. As science offered first new explanations and then new technologies, we have traded in the old contexts that informed human lives, bargaining them away in return for freedom, for liberation. The medieval church, which ordered Western civilization, gave way to more individualized religion; we read the Bible for ourselves, or not. Static peasant life, and guild life, in which the carpenter was the great-great-grandson of one carpenter, and the great-great-grandfather of another, gave way to the enormous dynamism of technology-driven capitalism: now 2 percent of Americans work as farmers, and our typical countryman changes jobs eight times in his life. Conservatives whimpered about the threat to order almost from the start—they *knew* Galileo was trouble, could sense the trajectory from him and his telescope to Nietzsche and the death of God. But radicals saw it just as clearly. Marx and Engels, of course, offered the single greatest description of this phenomenon when they wrote in 1848 that "All fixed, fast-frozen relations, with their train of ancient and venerable prejudices and opinions, are swept away; all new-formed ones become antiquated before they can ossify. *All that is solid melts into air*."[152]

What's really amazing about what Max Weber called the "disenchantment" of the world was how long it took. "After every announcement of the technological conflagration, much traditional and natural reality has remained to be consumed," writes the philosopher Albert Borgmann.[153] So, in the last century, the invention of the car offered the freedom of mobility, at the cost of giving up the small, coherent physical universes most people had inhabited. The invention of radio and television allowed the unlimited choices of a national or a global culture, but undermined the local life that had long persisted; the old people in my small rural town can still recall when "visiting" was the evening pastime, and how swiftly it disappeared in the 1950s, when CBS and NBC arrived. The 1960s seemed to mark the final rounds of this endless liberation: the invention of divorce as a mass phenomenon made clear that family no longer carried the meaning we'd long assumed, that it could be discarded as the village had been discarded; the pill and the sexual revolution freed us from the formerly inherent burdens of sex, but also often reduced it to the merely "casual."

Whether all this was "good" or "bad" is an impossible question, and a pointless one. These changes came upon us like the weather; "we" "chose" them only in the broadest sense of the words. They were upon us before we could do anything about them. You may keep the TV in the closet, but you still live in a TV society. The possibility of divorce now hovers over every marriage, leaving it subtly different from what it would have been before. What's important is that all these changes went in the same direction: they traded context for individual freedom. Maybe it's been a worthwhile bargain; without it, we wouldn't have the prosperity that marks life in the West, and all the things that prosperity implies. Longer life span, for instance; endless choice. But the costs have clearly been real, too: we've tried hard to fill the hole left when community disappeared, with "traditional values" and evangelical churches, with back-to-the-land communes and New Age rituals. But those frantic stirrings serve mostly to highlight our radical loneliness. Even the surrounding natural world, as I argued in *The End of Nature,* no longer serves as a ground, a context; we've reshaped it so thoroughly, now changing even its climate, that it reflects our habits and appetites and economies instead of offering us a doorway into a deeper world.

The past five hundred years have elevated us to the status of individuals, and reduced us to the status of individuals. At the end of the process, that's what we are—empowered, enabled, isolated, disconnected individuals. Call it blessing or call it curse or call it both, that's where we find ourselves. Our greatest cultural artifact is probably *Seinfeld,* a television program devoted to exploring what it means to live a life that has no context, that has no meaning. A show, famously, about nothing. The great danger, in other words, of the world that we have built is that it leaves us vulnerable to meaninglessness—to a world where consumption is all that happens, because there's nothing else left that means anything. In a way that once was unthinkable, we now have to ask ourselves, "Is my life amounting to something? Does it have weight and substance, or is it just running away into nothing, into something insubstantial?"[154] And the only real resource that many of us have against that meaninglessness, now that the church and the village and the family and even the natural world can't provide us with as much context as before, is our individual selves. We have to, somehow, produce all that context for ourselves; that's what a modern life is about. There's no use moaning about it; it may well be better than what came before. In any event, it's who we are, where we are, how we are, what we are, why we are. We've got to answer those questions pretty much on our own.

But now—and finally, here's the heart of the argument—*we stand on the edge of disappearing even as individuals*. Most of the backdrops have long since been dragged off the stage, and most of the other actors have mostly vanished; each of us is giving our existential monologue, trying to make it count for something. But in the wings, the genetic engineers stand poised to slip us off the stage as well, and in so doing to ring down the curtain on the entire show.

• • •

It doesn't seem so at first; if anything, just the opposite. The engineers promise to complete the process of liberation, to free us (or, rather, our offspring) from the limitations of our DNA, just as their predecessors freed us from the confines of the medieval worldview, or the local village, or the family. They can, they promise confidently, remove the ties that bind us—the genes that allow us to fall into ill health, or that keep us from being more intelligent, or more muscular, or more handsome, or happier. It seems as if, with their splicing and snipping, they want only to remove one more of the stones that weigh us down; that without it we will bound even higher, be more truly liberated.

In fact, though, whatever you think of the last five hundred years, this is one liberation too many. We are snipping the very last weight holding us to the ground, and when it's gone we will float silently away into the vacuum of meaninglessness.

What will you have done to your newborn when you have installed into the nucleus of every one of her billions of cells a purchased code that will pump out proteins designed to change her? You will have robbed her of the last possible chance of understanding her life. Say she finds herself, at the age of sixteen, unaccountably happy. Is it *her* being happy—finding, perhaps, the boy she will first love—or is it the corporate product inserted within her when she was a small nest of cells, an artificial chromosome now causing her body to produce more serotonin? Don't think she won't wonder: at sixteen a sensitive soul questions everything. But perhaps you've "increased her intelligence"—and perhaps that's why she is questioning so hard. She won't even be sure whether the questions are hers.

Here's Gregory Stock, the UCLA professor and outspoken proponent of this engineering, explaining how it will work: "People will be inclined to give their children those skills and traits that align with their own temperament and lifestyles. An optimist may feel so good about his optimism and energy that he wants more of it for his child. A concert pianist may see music as so integral to life that she wants to give her daughter greater talent than her own. A devout individual may want his child to be even more religious and resistant to temptation."[155] In other words, having managed, in many ways against the odds, to create a context (optimism, devotion, artistry) for their lives, parents will be able to pass it on. But what a poisoned gift. Scientists—"neurotheologians," someone has called them—have pinpointed the regions of the parietal lobe that quiet down when Catholic nuns and Buddhist monks pray.[156] Perhaps before long we will be able to amplify the reaction. As a result, the minister's son may be even more pious than he is—but if he has any brain left to himself, he will question that piety at the deepest level, wonder constantly whether it means anything or if it's so much (literal) brainwashing. And if he doesn't question it, if the gene transplant takes so deeply that he turns into an anchorite monk living deep in the desert, then his faith is utterly meaningless, far more meaningless than the one his medieval ancestor inherited by birthright. It would be a faith literally beyond questioning, and hence no faith at all. He would be, for all intents and purposes, a robot.

And the piano player's daughter? A player piano as much as a human, doomed to create a particular context for herself, ever uncertain whether it is her skill and devotion or her catalogue proteins that move her fingers so nimbly, her music soured before it is made. Because the point was never the music itself; the inclination and then the effort were what created the meaning for her mother. If she injects all that into her daughter's cells, she robs her forever of the chance to make music her own authentic context—or to choose something else (dance, art, cooking) as the act that brings her life to life.

I began by talking about running, because it is one of the contexts I've created for myself, one of the things that orders my life, fills it with metaphor and meaning. If my parents had somehow altered my body so that I could run more quickly, that fact would, as I said, have robbed it of precisely that meaning I draw from it. The point of running, for me, is not to cover ground more quickly; for that, I could use a motorcycle. The point has to do with seeking out my limits, centering my attention: finding out who I am. But that's very difficult if my body has been altered, if the "I" and the Sweatworks2010 GenePack are entwined in the twists of the double helix. And if my mind has been engineered to make me want to push through the pain of running, or not notice it at all, then the point has truly vanished. My effort to carve out some context for myself is in vain; I might as well be Seinfeld, sitting on his couch and cracking wise about the pointlessness of it all.

And when it comes time for me to visit the clinic and program my own offspring, how do I know why I choose what I choose? Most likely, to quote Stock once more, "enhancements of this sort by parents [will] engender mindsets disinclined to attenuate the traits in their own children," and so "such traits may reinforce themselves from generation to generation and push the limits of genetic possibility and technical know-how."[157] Because, that is, one late-twentieth-century woman found solace and meaning in playing the piano, her descendants yea unto the generations are condemned to an ever-deepening spiral of musicality, one that they did not choose and that may haunt them, depending on how much consciousness remains, with the question of why exactly they feel so compelled to compose.

We flirt, of course, with these possibilities already. When taken by people who are not in obvious medical need, drugs like Prozac may smooth out identity, stunt emotional growth; at the very least, as many have noted, they raise the question of how you tell who you really are. But Prozac and its soma sisters remain, for the moment, pills. They are designed to help people through bad patches. You can refuse to take them, you can stop taking them; they are not *you* in the sense that they would be if municipal officials loaded them into the water supply. And certainly not in the way they would be you if your optimistic father had determined that he wanted double grande optimism in his son and so worked the extra serotonin into your very wiring, syringed it in as an ineradicable tattoo.

Some of these changes may well make us more comfortable. Stock again: "If we had the power to protect our future child, we might be very reluctant to leave him or her with a predisposition for recurring bouts of dark depression." Not even "the knowledge that our child might use these distressing periods to good purpose" would "make our decision to forgo germinal interventions any easier. I suspect that most parents would make the safe choices and avoid the ragged uncertainties at the edges of human possibility."[158] Certainly they would after a few generations, when the fear of a passage they had never known would make parents-to-be doubly wary. But in that increasing suburbanization of our being, the chance for emotional growth, for becoming "real" in some deep sense, would dwindle ever further.

We will explore in greater depth the shape of this meaninglessness, try to chart the emotional geography of this world we are so near to stepping into. But for the moment, bear this thought in mind as a touchstone: we may be the last generations able even to undertake such an exercise. In the words of Richard Hayes, one of the leading crusaders against germline manipulation: "Suppose you've been genetically engineered by your parents to have what they consider enhanced reasoning ability and other cognitive skills. How could you evaluate whether or not what was done to you was a good thing? How could you think about what it would be like *not* to have genetically engineered thoughts?"[159]

If the programming works, in other words, then you will have turned your child into an automaton of one degree or another; and if it only sort of works, you will have seeded the ground for a harvest of neurosis and self-doubt we can barely begin to imagine. If "Who am I?" is the quintessential modern question, you will have guaranteed that your offspring will never be able to fashion a workable answer.

Beginning in the 1960s, an émigré psychologist named Mihaly Csikszentmihalyi began to study the experience of enjoyment, a research project that helps show why genetic enhancement won't make us any happier. "During my thesis research, as I watched and photographed painters at their easels, one of the things that struck me most vividly was the almost trance-like state they entered when the work was going well. . . . The motivation to go on painting was so intense that fatigue, hunger, or discomfort ceased to matter. Why?" Not, he decided, for the reasons a behaviorist might have suggested—that they wanted the reward of a finished painting. In fact, "I noticed that the artists I was observing almost immediately lost interest in the canvas they had just painted. Typically they turned the finished canvas around and stacked it against a wall. Nor were they particularly eager to show it off, or very hopeful about selling it. They could hardly wait to start on a new one."[160]

Finding out why turned out to be a problem worthy of a life's work. Soon Csikszentmihalyi was studying a wide variety of people who found deep joy in some particular activity—rock climbing, chess, basketball, composing music—and trying to understand why their absorption was so satisfying. Psychoanalysis offered lousy answers, positing that people played chess in order to "vicariously kill their father (the opponent's king) and thus release the oedipal tension which frustrates them in real life" or that they climbed rock spires out of a "sublimated penis worship" that implied latent homosexuality.[161] Instead, Csikszentmihalyi's surveys soon revealed that "the most basic requirement" of such activities "is to provide a clear set of challenges," "a going beyond the known, a stretching of one's self toward new dimensions of skill and competence." For the basketball players he studied, the test came in competition. "A more subtle way of establishing physical competence is by matching one's skill against a physical obstacle or against the boundaries of one's own competence. Rock climbers take this option. Another way of going beyond the given is by confronting intellectual problems. Again, one can do so competitively, as chess players do, or by struggling against internal obstacles, as composers do."[162]

Csikszentmihalyi labeled this state of joyful absorption "flow," and found that it lay between the boredom of much of normal modern life, and the anxiety that often intrudes on that boredom. A person in a state of flow has neither more nor less challenge than she can handle. This was, for example, the state in which I found myself as the marathon wound to its finish, concentrating so fully on the task at hand that nothing else could enter my consciousness—the same feeling I get, occasionally, when I am writing something of real worth. Csikszentmihalyi quotes rock climbers ("You are so involved in what you are doing that you aren't thinking of yourself as separate from that immediate activity"), modern dancers ("Your concentration is very complete. Your mind isn't wandering; you are totally involved in what you are doing"), and chess players ("The concentration is like breathing—you never think of it. The roof could fall in and, if it missed you, you would be unaware of it").[163] Flow, he found, occurred only along a fairly narrow pathway between too little challenge (which led to boredom) and too much (which led to anxiety).

It's clear, then, that souping someone up genetically will not increase his opportunities to fall into this flow state. It may make him "better," but "better" is not the point—he can already max out his enjoyment by climbing a route up the rocks that tests him fully. If he were genetically altered to have stronger fingers and forearms, he would be able to climb harder routes—but he wouldn't go "deeper" into the flow state. (And perhaps all the routes in the neighborhood would now be too easy to offer the necessary challenge; he'd have to fly to Yosemite in search of the joy he once found on a nearby face.) The chess player whose memory has been upgraded will still be able to find competition—his ranking will go up, and perhaps there will be a special league for people with his brand of DNA. (If his upgrade is too good, however, he may run out of opponents. And, as Csikszentmihalyi notes, "many chess champions have been known to suddenly to go to pieces once they reached the top and were deprived of the accustomed challenges."[164]) But there will be no *extra* satisfaction for him—the joy comes not from excelling against some arbitrary standard, but from excelling against whatever your limits happen to be. The girl scrambling up a boulder, if it's at the edge of her possible, is as fully engaged as the veteran climber working her way up Half Dome.

So an upgrade won't multiply your joy. Instead, it might well sap joy, because forgetting the self seems to be a key part of falling into the flow. "When things are going poorly, you start thinking about yourself," says one rock climber. Explains another, "If you can imagine yourself becoming as clear as when you focus a pair of binoculars, everything's blurred and then the scene becomes clear as you focus them. If you focus yourself in the same way, until all of you is clear, you don't think about how you're going to do it, you just do it."[165] Imagine how easy it would be to fog up that clarity. What if you were thinking, in the back of your head, "Is it really me doing this? Is it my programming? Am I losing myself, or is that feeling merely an artifact of my engineering?" And those are precisely the sorts of thoughts that would rise in your mind—because, in some ways, *the whole point* of flow experiences is to know yourself better. In the words of one of Csikszentmihalyi's rock climbers, up on the face "you see who the hell you really are. It's important to learn about yourself, to open doors into the self. . . . The self-consciousness of society is like a mask. We are born to wear it. Up there you have the greatest chance of finding your potential for any form of learning. Up there the false masks, costumes, personae that the world puts on you—false self-consciousness, false self-awareness—falls away."[166] Knowing that you were *constructed,* that your very proteins may have been designed to incline you toward thrill-seeking (or toward safety) pollutes that clarity. Why would you want to add a new, complicating layer of self-doubt, one that could undermine precisely those activities we've devised to let us make life meaningful? "Many of the people we interviewed," writes Csikszentmihalyi, "especially those who most enjoy whatever they are doing, mentioned that at the height of their involvement with the activity they lose a sense of themselves as separate entities, and feel a harmony and even a merging of identity with the environment."[167] But how can you become one with the rock if you're two with yourself? (And if you're *not* two with yourself—if the treatment has worked so smoothly that your various choices never cross your mind—then you've become a species of automaton. In which case, why are you trying to "find yourself" out there on the cliff face? Just refer to your design specs.)

If there's no opportunity for joy, suggests Csikszentmihalyi—if we never find some activity that leads us into flow, or if that flow is shut off by spiraling self-doubt—then we're likely to settle for pleasure instead: for the sterile comforts of the television or the bottle or the slot machine, which offer no challenge, no room for growth, but only a kind of vacation from boredom or from worry. Or we will "work harder for extrinsic rewards, to accumulate some tangible feedback for our existence. Status, power, and money are signs that one is competent, that one is acquiring control. But these are secondary rewards that matter only when the primary enjoyment is not available."[168] We can see all this displayed in the world around us, of course; we can feel it in ourselves. Joy is elusive enough already, and its various substitutes too much with us.

Albert Borgmann has spent his career trying to do as a philosopher what Csikszentmihalyi did as a psychologist: figure out what constitutes the good life. He argues that although the technology that has driven the change of the last five hundred years has done much good—freed us from darkness, cold, heat, and hunger—the cost has been to confuse us into thinking that technology is always the answer, that it enriches our lives in some deep way as easily as it makes them more comfortable and convenient. That a wide-screen TV will make a radical difference. And hence we increasingly lead lives of passive consumption. Borgman calls this the "device paradigm": when devices "fill our lives we are reduced to disengaged consumers of the commodities these devices provide."[169]

And so he, too, argues for challenge, what he calls "the practice of engagement." Find a "focal practice," he argues, and use the time freed up for us by technology—the fact that we can buy our food at a supermarket, say—to go so deeply into running, or cooking, or music making that it "orders and orients" our existence. This is not easy: "on the spur of the moment we normally act out what has been nurtured in our daily practices as they have been shaped by the norms of our time."[170] That is to say, we are likeliest to switch on *Law and Order,* hop onto the Net, fire up our copy of Doom. But every time we do, we leave behind the possibility of a reality that "gathers and illuminates our world," substituting instead "the insubstantial and disconnected glamour" of the technological world, which "provokes disorientation and distraction." How do *you* feel after a full evening of watching prime time? Like maybe you wished you'd done something else?

And now we seem bent on making our children into devices. At the moment we still perceive—sometimes dimly—the distinction between the real and the artificial. In fact, there are signs that more of us take it more seriously all the time: worried and unsatisfied by a life in front of the TV, we reach for some of these more fulfilling disciplines. We run; we take up instruments. Or at least, we yearn to do so. But how fulfilling will those focal practices be once the line is fully blurred—once we have been turned into a kind of tool? If you've been designed and programmed to run, what meaning can running hold? It becomes an endless round on a treadmill, except that the treadmill is inside you—you take it out in the woods when you go for a trail run, and onto the beach when you run beside the breakers. Making devices of ourselves would be the logical end of our technological momentum; it would end the tension we feel between the real and the artificial. And that might be a relief; there'd be nothing nagging at us to go "make more" of our lives. But that tension is the last remaining fully human part of us, and to give it up is like letting go of the last vine in the jungle; without it we free-fall toward who knows where.

Beginning the hour my daughter came home from the hospital, I spent part of every day with her in the woods out back of our house, showing her trees and ferns and chipmunks and frogs. One of her very first words was "birch," and you couldn't have asked for a prouder papa. She got her middle name from the mountain we see out the window; for her fifth birthday she got her own child-sized canoe; her school wardrobe may not be relentlessly up-to-date, but she's never lacked for hiking boots. As I write these words, she's spending her first summer at sleep-away camp, one that we chose because the kids sleep in tents and spend days in the mountains. All of which is to say that I have done everything in my power to try to mold her into a lover of the natural world. That is where my deepest satisfactions lie, and I want the same for her. It seems benign enough, but it has its drawbacks: it means less time and money and energy for trips to the city and music lessons and so forth. As time goes on and she develops stronger opinions of her own, I yield more and more, but I keep trying to stack the deck, to nudge her in the direction that's meant something to me. On a Saturday morning, when the question comes up of what to do, the very first words out of my mouth always involve yet another hike. I can't help myself.

In other words, we already "engineer" our offspring in some sense: we do our best, and often our worst, to steer them in particular directions. And our worst can be pretty bad. We all know people whose lives have been blighted trying to meet the expectations of their parents. We've all seen the crazed devotion to getting kids into the right schools, the right professions, the right income brackets. Parents try to pass down their prejudices, their politics, their attitude toward the world ("We've got to toughen that kid up—he's going to get walked all over"). There are fathers who start teaching the curveball at the age of four, and sons who are made to feel worthless if they don't make the Little League traveling team. People move their house so that their kids can grow up with the right kind of schoolmates. They threaten to disown them for marrying African Americans, or for not marrying African Americans. No dictator anywhere has ever tried to rule his subjects with as much attention to detail as the average modern parent.

Why not take this just one small step further? Why not engineer children to up the odds that all that nudging will stick? In the words of Lee Silver, the Princeton geneticist, "Why not seize this power? Why not control what has been left to chance in the past? Indeed, we control all other aspects of our children's lives and identities through powerful social and environmental influences. . . . On what basis can we reject positive genetic influences on a person's essence when we accept the rights of parents to benefit their children in every other way?"[171] If you can buy your kid three years at Deerfield, four at Harvard, and three more at Harvard Law, why shouldn't you be able to turbocharge his IQ a bit?

But most likely the answer has already occurred to you as well. Because you know plenty of people who managed to rebel successfully against whatever agenda their parents laid out for them, or who took that agenda and bent it to fit their own particular personality. In our society that's often what growing up is all about—the sometimes excruciatingly difficult, frequently liberating break with the expectations of your parents. The decision to join the Peace Corps (or the decision to leave the commune where you grew up and go to business school). The discovery that you were happiest davening in an Orthodox shul three hours a day, much to the consternation of your good suburban parents, who almost always made it to Yom Kippur services; the decision that, much as you respected the Southern Baptist piety of your parents, the Bible won't be your watchword.

Without the grounding offered by tradition, the search for the "authentic you" can be hard; our generations contain the first people who routinely shop for religions, for instance. But the sometimes poignant difficulty of finding yourself merely underscores how essential it is. Silver says the costs of germline engineering and a college education might be roughly comparable; in both cases, he goes on, the point is to "increase the chances the child will become wiser in some way, and better able to achieve success and happiness."[172] But that's half the story, at best. College is where you go to be exposed to a thousand *new* influences, ideas that should be able to take you in almost any direction. It's where you go to get out from under your parents' thumb, to find out that you actually don't have to go to law school if you don't want to. As often as not, the harder parents try to wrench their kids in one direction, the harder those kids eventually fight to determine their own destiny. I am as prepared as I can be for the possibility—the probability—that Sophie will decide she wants to live her life in the concrete heart of Manhattan. It's *her* life (and perhaps her kids will have a secret desire to come wander in the woods with me).

We try to shape the lives of our kids—to "improve" their lives, as we would measure improvement—but our gravity is usually weak enough that kids can break out of it if and when they need to. (When it isn't, when parents manage to bend their children to the point of breaking, we think of them as monstrous.) "Many of the most creative and valuable human lives are the result of particularly difficult struggles" against expectation and influence, writes the legal scholar Martha Nussbaum.[173]

That's not how a genetic engineer thinks of his product. He works to ensure absolute success. Last spring an Israeli researcher announced that he had managed to produce a featherless chicken. This constituted an improvement, to his mind, because "it will be cheaper to produce since its lack of feathers means there is no need to pluck it before it hits the shelves." Also, poultry farmers would no longer have to ventilate their vast barns to keep their birds from overheating. "Feathers are a waste," the scientist explained. "The chickens are using feed to produce something that has to be dumped, and the farmers have to waste electricity to overcome that fact."[174] Now, that engineer was not trying to influence his chickens to shed their feathers because they'd be happier and the farmer would be happier and everyone would be happier. He was inserting a gene that created a protein that made good and certain they would not be producing feathers. Just substitute, say, an even temperament for feathers, and you'll know what the human engineers envision. "With reprogenetics," writes Silver, "parents can gain *complete control* [italics mine] over their destiny, with the ability to guide and enhance the characteristics of their children, and their children's children as well."[175] Such parents would not be calling up their children on the phone at annoyingly frequent intervals to suggest that it's time to get a real job; instead, just like the chicken guy, they would be inserting genes that produced proteins that would make their child behave in certain ways throughout his life. You cannot rebel against the production of that protein. Perhaps you can still do everything in your power to defeat the wishes of your parents, but that protein will nonetheless be pumped relentlessly into your system, defining who you are. You won't grow feathers, no matter how much you want them. And maybe they can engineer your mood enough that your lack of plumage won't even cross your mind.

Such children will, in effect, be assigned a goal by their programmers: "intelligence," "even temper," "athleticism." (As with chickens, the market will doubtless lean in the direction of efficiency. It may be hard to find genes for, say, dreaminess.) Now two possibilities arise. Perhaps the programming doesn't work very well, and your kid spells poorly, or turns moody, or can't hit the inside fastball. In the present world, you just tell yourself that that's who he is. But in the coming world, he'll be, in essence, a defective product. Do you still accept him unconditionally? Why? If your new Jetta got thirty miles to the gallon instead of the forty it was designed to get, you'd take it back. If necessary, you'd sue. You'd call it a lemon.

Or what if the engineering worked pretty well, but you decided, too late, that you'd picked the wrong package, hadn't gotten the best features? Would you feel buyer's remorse if the kid next door had a better ear, a stronger arm?

Say the gene work went a little awry and left you with a kid who had some serious problems; what kind of guilt would that leave you with? Remember, this is not a child created by the random interaction of your genes with those of your partner—this is a child created with specific intent. Does *Consumer Reports* start rating the various biotech offerings?

What if you had a second child five years after the first, and by that time the upgrades were undeniably improved: How would you feel about the first kid? How would he feel about his new brother, the latest model?

The other outcome—that the genetic engineering works just as you had hoped—seems at least as bad. Now your child is a product. You can take precisely as much pride in her achievements as you take in the achievements of your dishwashing detergent. It was designed to produce streak-free glassware, and she was designed to be sweet-tempered, social, and smart. And what can she take pride in? Her good grades? She may have worked hard, but she'll always know that she was specced for good grades. Her kindness to others? Well, yes, it's good to be kind—but perhaps it's not much of an accomplishment once the various genes with some link to sociability have been catalogued and manipulated. Here's an analogy, a somewhat loaded one: in recent years many critics of affirmative action have claimed it can undermine the self-esteem of its beneficiaries, who are left to wonder, "Am I here at Yale because my skin's a certain color, or because I'm smart?" I don't know how widespread this feeling is—after all, most of the people helped by affirmative action have already overcome great odds and have much to take pride in. But I have no doubt that these qualms would be one of the powerful psychological afflictions of the future—at least, until someone figures out a fix that keeps the next generations from having such bad thoughts.

Britain's chief rabbi, Jonathan Sacks, was asked a few years ago about the announcement that Italian doctors were trying to clone humans. "If there is a mystery at the heart of human condition, it is otherness: the otherness of man and woman, parent and child. It is the space we make for otherness that makes love something other than narcissism."[176] I remember so well the feeling of walking into the maternity ward with Sue, and walking out with Sue and Sophie: where there had been two, there were now, somehow, three, each of us our own person, but now commanded to make a family, a place where we all could thrive. She was so mysterious, that Sophie, and in many ways she still is. There are times when, like every parent, I see myself reflected in her, and times when I wonder whether she's even related. She's ours to nurture and protect, but she is who *she* is. That's the mystery and the glory of any child.

"Mystery," however, is not one of the words that thrills engineers. They try to deliver solid bridges, unyielding dams, reliable cars. We wouldn't want it any other way. The only question is whether their product line should be expanded to include children.

The parent as god-king, laying down the channel through which his child's life must flow: that's one image from this new world. In the words of Leon Kass, "any child whose being, character, and capacities exist owing to human design does not stand on the same plane as its makers. As with any product of our making, no matter how excellent, the artificer stands above it, not as an equal but as a superior, transcending it by his will and creative prowess."[177]

But here's a twist to bear in mind. If the engineering works as intended, the offspring will be superior to their parents. With a higher IQ, or a more manageable temper, or a better ear, or quicker reflexes. Not "better" as when a son grows in strength while his father declines, till one day their positions are reversed, but categorically better, of a higher order. *Different*. One reason we love and nurture our kids, or so the biologists tell us, is from an inarticulate desire to pass along our genes. But these won't be our genes precisely; they'll belong as much to whatever multinational created them. And these kids won't be our kids, not exactly. The gulf between their generation and ours will be enormous, their "evolution" accelerated. If the engineering all works as it's supposed to, they will be creating a world in which they are comfortable, and not their parents. That happens now, of course; it's beyond trite to observe that your seven-year-old is better able than you to program the VCR or download files from the Web. Sometimes it happens in even more exaggerated form, as when the children of first-generation immigrants must take over the adults' role in dealing with the English-speaking world. But now that difference will be physical, encoded. Children will in some sense be of a different species, or at least a different strain.

That will cause confusion aplenty: talk about undermining parental authority. ("Dad, you just don't understand" will have a different, more literal meaning.) But it will also, with each progressive generation, sever those children more fully from their human past.

Right now both our genes and the limits that they set on us connect us with every human that came before: Our fight-or-flight response, ingrained since the days when it helped us survive in a dangerous world. My love of running, somehow connecting me to that moment six million years ago when my kind ventured out of the forests and onto the savannas, discovering that bipedal locomotion aided greatly in the hunt. (By some theories, it evolved to help us thermoregulate; the upright body absorbed 60 percent less solar energy than the same ape-man on all fours.)[178] And connecting me to the moment when Pheidippides sank dying to his knees with news of the victory over the Persians. Twenty-six miles is *still* hard, though it may not be for long.

Human beings can look at rock art carved into African cliffs and French caves 30,000 years ago and feel an electric, immediate kinship. We are, by and large, the same people, more closely genetically related to one another than we may be to our engineered grandchildren. We've gone from digging sticks to combines, and from drum circles to symphony orchestras (and back again to drum circles), but we still hear in the same range and see in the same spectrum, still produce adrenaline and dopamine in the same ways, still think in many of the same patterns. And if these patterns sometimes cause us trouble (group solidarity, a focus on short-term danger at the expense of long-term trouble) they show, nonetheless, that we are the same people.

From the sublime to the (almost) ridiculous. Consider for a minute the American religion of baseball, whose record book links past to present to future. "Baseball's time is seamless and invisible, a bubble in which players move at exactly the same pace and rhythms as their predecessors," writes Roger Angell, the game's greatest chronicler. "This is the way the game was played in our youth, and our father's youth, and even back then—back in the country days—there must have been the same feeling that time could be stopped."[179] One of my earliest sports memories dates to 1967, when my father called me over to the TV to watch Carl Yastrzemski up at bat for the Red Sox in his glorious Triple Crown season, when he became the last man to lead the league in homers, RBIs, and batting average. "That's one of the greatest players of all time," my father said. "Remember him." Remember him I did. And I made sure to take Sophie, who was five at the time, out to see Mark McGwire in his record year.

But a few weeks after we went to that game, some reporter noticed a vial of androstenedione, a steroidlike substance, in McGwire's locker. By last year, as home run records kept falling steadily, people began to loudly complain that the entire game had become drug-infested. Part of the concern was for the players themselves: suddenly they were swinging so hard that they literally tore their bodies apart. "You wouldn't believe the Achilles tendon ruptures, the quadriceps ruptures, the hamstring tears, the massive rotator cuff tears, the tearing of the biceps muscles at the elbow joints," said one sports orthopedist. "There's just too much mass for the body to handle."[180] But in truth, more fans mourned for that connection with the past. "The lifeblood of baseball is statistics, numbers, and records, which fans take on faith since they will only see an infinitesimal fraction of the actual games," wrote one indignant sportswriter. "What happens when fans no longer accept the numbers as a true reflection of the players' on-field performances?"[181] What happens is, the fans no longer care. Home runs become mere spectacle; the bond with what went before is broken. "Ted Williams hit 521 home runs at 6'3" and not much more than 180 pounds," wrote the *Sports Illustrated* columnist Rick Reilly. "Williams would look like Poindexter the Stickboy in a clubhouse today. This isn't baseball. This is my-test-tube-can-beat-up-your-test-tube. . . . Hall of Fame greats such as Carl Yastrzemski will soon be passed on the home run list by puffed-up one-trick freaks who couldn't have scrubbed their jocks."[182]

Baseball now promises to clean up its act by starting to test for drugs, just like every other sport. Even if it doesn't, we will survive the loss of baseball purity, though the world will be a little diminished by it. But say we move into an era where this kind of improvement happens constantly, and with every human attribute. Connection starts to vanish. One year not long ago, John McCardell, the president of Middlebury College, addressed the arriving students just before a photographer snapped their freshman class. "Think of students yet unborn gazing upon your countenances," he said. "They may laugh at your choice of dress and hairstyle. But if they look into your faces, they may also, if only fleetingly, see into your hearts, because they will, in fact, be staring into a mirror. Their hopes and dreams will be no different from yours."[183] Always before, that's been true. But the class of 2050 may be very different indeed. Perhaps so different that when they sit down to read a book from our time it will be of historical interest only, the record of a different creature. "I believe in the continuousness of human existence," declared the writer Richard Rodriguez. "I renew that faith when I read a book that was written three hundred years ago or two thousand years ago, and I still relate to it. I believe our human experience has that kind of durability."[184] We still have Macbeth's ambition, Lear's neediness running through our veins. We have the same proteins in roughly the same amounts. But for how much longer?

It's already a little hard to get your kids interested in their great-grandparents, who are far enough back to be out of focus, blurry. But what if the physical line that connects you to your own ancestors were strained or broken, what if your inheritance from them were only an echo of some nurture, not a very real chunk of nature? Adopted children already deal with this question, of course, but they know that somewhere back there a genetic link exists—witness their often ferocious attempts to sleuth out their biological relations. What if the day comes when no one has a "birth mother," not anywhere? Here's Dr. Michael West, the CEO of Advanced Cell Technology, imagining a future when he can engage in "chromosome shuttling by microsome-mediated transfers in, say, skin cells," which would mean "I could take a Y chromosome from Arnold Schwarzenegger, a Chromosome X from Bob Barker, a Chromosome 6 from Robin Cook, and so on, and assemble a human being with 46 parents, all male for that matter—a child with 46 fathers and no mother."[185] You don't have to wait that long, however, or dream that ambitiously. (Bob Barker!) The first child whose genes come at least in part from some corporate lab, the first child who has been "enhanced" from what came before—that's the first child who will glance back over his shoulder and see a gap between himself and human history.

But here's the really awful part: he won't be able to look forward, either. He won't be able to imagine himself connected with those who will come after him. Because, of course, by then there will be better upgrades. They'll be Windows 2050 to his Atari. He'll be marooned forever on his own small island, as will all who follow him.

Though our lives in the developed world are easy enough by comparison to lives in other places and other eras, challenges remain. Or at least, as when we run marathons, we can invent them. We are able, that is, to spend our lives at the basic human task of figuring out who we are. Our parents try to draw us maps, which we can follow slavishly, burn in the fires of our rebellion, or glance at from time to time as we chart our own courses. But these new technologies show us that human meaning dangles by a far thinner thread than we had thought. What if the ending to our story has already been written, our compass already set? What if we have been programmed, or at least must suspect each time we choose a path that we have been nudged in that direction by our engineered cells? Who then *are we*?

CHAPTER TWO

Even More

On the same day in November 2001 that Advanced Cell Technology announced it had cloned the first human embryo, a group of Israeli scientists made an almost equally stunning declaration. They had used biological molecules to create a tiny, programmable computer—so tiny that a trillion of them could "co-exist and compute in parallel, in a drop the size of 1/10 of a milliliter of watery solution held at room temperature." The computer hardware consisted of naturally occurring enzymes that manipulate DNA; it could be programmed to perform simple tasks by choosing particular software molecules to be mixed in solution.[1] A month later the biggest American chipmakers, Intel and AMD, again on the same day, announced that they had both managed to radically speed up their new-generation microprocessors by reducing the "gate length," or space between two key components on a transistor. They had it down to 15 nanometers, or about the width of sixty atoms. By September 2002, Hewlett-Packard was announcing that it could space wires ten nanometers apart on its chips.[2] Three weeks past that, in October 2002, IBM scientists working with carbon monoxide molecules on a copper surface managed to build a logic circuit 1/260,000th the size of the current silicon models—a circuit that covered less than a trillionth of a square inch.[3]

The biggest clichés often hide deep truth in plain view, and this is one of those cases. We've said so often that the world is speeding up that it's become a truism, a commonplace. But it is the central fact of our time. We live in an era of exponential growth. Our computers don't just get faster, they double in speed every year or two, becoming twice as powerful, and then four times, and then . . . Which means that, close on the heels of genetic engineering, other technologies are now appearing that only a few years ago seemed more like science fiction than science. Techniques such as advanced robotics and nanotechnology simply must be taken seriously, because on their own, and in combination with genetic engineering, they could quickly evaporate human meaning.

"Quickly" is the key word. It's the new nature of time—its radical compression—that we first need to comprehend if we're going to make sense of these technologies. To understand this change, consider the careers of three of the great inventors of our era.

Rodney Brooks grew up in Adelaide, South Australia, in the 1950s—a remote city in an era when remoteness still meant something. "It was an isolated place at the end of the technological earth," he recalls. "I grew up a nerd in a place that did not know what a nerd was. I would stare through the window at the IBM mainframe in the financial district of the city, lusting after the technology."[4]

Hans Moravec was reared in Austria, where when he was four his father, an electrical engineer, gave him a wooden Erector set. He used it to build a model of a little man who would dance and wave his arms and legs when you turned a crank. "It excited me, because at that moment I saw you could assemble a few parts and end up with something more," he said.[5] "That's not a man. But it *acts* like a man."[6]

Ray Kurzweil grew up in New York, where by the age of twelve he was rooting through the bins outside the cut-rate electronics stores on Canal Street, salvaging parts to build early computers. In 1965, when he was in high school, he constructed a machine able to compose "classical" music; it won him the Westinghouse science award, and also a spot on *I've Got a Secret.*[7]

In the normal course of things, in the time we are used to living in, these men and thousands more like them—driven, brilliant, splendidly absorbed—would simply fill the same niche as those prodigies who came before. They'd be our new Edisons and Fords, our new Orville-and-Wilburs. They'd invent the next in the seemingly endless series of machines that for five hundred years have altered (for better and for worse) the ways we live. They'd push us slightly forward. When they were done with the planet, it would be different. But it wouldn't be a different planet.

Now, however, people like Kurzweil, Brooks, and Moravec imagine nothing less than the replacement of our species with some ill-defined successor race. "The emergence in the early twenty-first century of a new form of intelligence on Earth that can compete with, and ultimately significantly exceed, human intelligence will be a development of greater import than any of the events that have shaped human history," says Kurzweil.[8] "Those of us alive today, over the course of our lifetimes, will morph ourselves into machines," says Brooks.[9] "Biological species almost never survive encounters with superior species,"[10] says Moravec. Which is not something it would have occurred to Alexander Graham Bell to say.

They are so sure of themselves, at heart, because of Moore's law, a defining principle for our time. Gordon Moore, an inventor of the integrated circuit, who went on to run Intel, noted early on that the surface area of a transistor was shrinking by half about every two years—that, in other words, "every two years you can get twice as much circuitry running at twice the speed for the same price."[11] In fact, says Kurzweil, that same exponential curve predates the integrated circuit; if you go back to the card-based technology used to conduct the 1890 census and track computing through the wartime code-cracking machinery and the first vacuum-tube computers of the 1950s, you find the same degree of exponential growth.[12] It's just that for a long time no one noticed, because the growth began from such a small base. Moravec once tried to plot the pace of computer development against the speed with which biological evolution developed the human nervous system. The wormlike animals of the Cambrian, 540 million years ago, had a few hundred neurons, he pointed out, much like the first electromechanical computers of the 1940s with their "few hundred bits of telephone relay storage." The computers of 1955, with 100,000 bits of rotating memory, compare with the earliest fish. And so on, through the small mammals and the dinosaurs, until about the year 2000, when some high-powered personal computers boasted the 20 billion bits of RAM that were standard equipment for your average hominid ape of 30 million years ago. "We seem to be re-evolving mind (in a fashion) at ten million times the original speed."[13]

The punch line of this story is that we just happen to be alive at the brief and interesting moment when this growth starts to really matter—when it spikes. "It is in the nature of exponential growth that events develop extremely slowly for extremely long periods of time," writes Kurzweil, "but as one glides through the knee of the curve, events erupt at an increasingly furious pace. And that is what we will experience as we enter the twenty-first century."[14] *In the knee of the curve:* that's where things happen joltingly fast. Consider, says the science writer Damien Broderick, the way human travel has accelerated over the millennia. Graduating from walking to using donkeys to riding horses took many thousands of years; two centuries ago, steam trains started beating horses; then, in increasingly rapid succession, came cars and prop planes and jets. By 1953, he says, not even the Air Force technologists could believe what the trend curves were telling them: that within four years they would have achieved speeds great enough to lift payloads into orbit. But of course, *Sputnik* went into orbit right on schedule in 1957, and twelve years later Americans were bouncing around on the moon.[15]

At some point in the curve, this growth starts to go faster than human beings alone can manage. Machines start designing new machines; time compresses further. Whereas advances in integrated circuitry circa 1970 required teams of draftsmen working by hand on plastic sheets, "today's computer chips contain tens of millions of components, placed by design programs running on older computers."[16] Indeed, the rate of exponential growth itself seems to be exponentially growing. When the Human Genome Project was first proposed, critics said it would take 10,000 years; its backers said they'd be done by 2010. The project was finished early in 2000, "because DNA scanning technology grew at a double exponential rate."[17] Instead of millennia, or centuries, or decades between big developments, we now see a month, a week, a day. In the words of Bill Joy, the chief scientist at Sun Microsystems, "forget fiction, read the newspaper."[18] As the *New York Times* put it recently, "Moore's law is becoming less of an axiom and more of a drag race."[19]

A few technologists dare to point out that while the emperor has a great many bits, he also crashes regularly and needs to be rebooted. In the words of the virtual reality pioneer Jaron Lanier, "we don't seem to be able to write software much better as computers get much faster." If anything, "there's a reverse Moore's law observable in software: As processors become faster and memory becomes cheaper, software becomes correspondingly slower and more bloated, using up all available resources."[20] But for every cautionary word, a dozen disciples preach nanotube computing, optical computing, DNA computing, crystalline computing, quantum computing: enough options to "keep the law of Accelerating Returns going in the world of computation for a very long time."[21] More than long enough to reach the point where computers have more connections, and process information more rapidly, than the human brain. In fact, by most estimates we're excruciatingly near the point of the curve where silicon passes flesh. Our brains possess something like 100 billion neurons, and by most calculations 100 billion bits of RAM will be standard equipment in computers within half a decade.

On its own, of course, all that extra computer power just means quicker calculations; to really change the world, for better or for worse, the power must be translated into new technologies. We will examine the two most likely possibilities—first robotics and then nanotechnology—because they are far enough along in the development curve so that we can at least begin to assess their possibilities. And because they hold the potential to end life as we have known it, in all the senses of that phrase.

At the moment, says Hans Moravec, "protection, repair, cleaning, transport and so forth . . . remain in human hands."[22] We are still necessary, at least to ourselves. This, it must be said, is not for lack of dreaming; people tinkered with automatons for centuries before anyone even imagined DNA. Pierre Jacquet-Droz, a Frenchman enthralled by intricate clockwork mechanisms, managed as early as 1768 to build a robot he called the scribe, which could be programmed to write any message of forty characters or less. "He moves in an uncannily realistic manner, pausing periodicially to dip his pen into the inkwell and shaking off any excess ink." The same workshop produced a robotic draftsman who could draw profile sketches of Louis XV (and bow his head to blow away the lead dust left on the page by his pencil), and Marianne the Musician, "a charming young woman who actually plays a tiny organ with all ten fingers. . . . Her breast rises and falls in an uncanny simulation of restrained breathing."[23]

Robots remained toys for the next couple of hundred years, right through the industrial revolution. We got good at making machinery, which eliminated the need for plenty of human workers. But only with the advent of microelectronics and computers did people begin to dream realistically of machines that would *be* workers, as flexible and perhaps eventually as smart as the humans who created them.

As early as the 1950s, writes Moravec, "it was common opinion in the Artificial Intelligence labs that, with the right program, readily available computers could encompass any human skill"; after all, computers were already calculating electric bills and mailing out Social Security checks, doing the work of thousands of calculators. And so, at universities and research centers around the country and the globe, teams of researchers attempted to deliver on the promise of AI—to build robots that could think. At first these researchers wore white shirts and dark ties and carried slide rules; fifteen years later they wore jeans and had long hair; but they were all engaged in the mostly fruitless battle to teach, say, small wheeled machines to navigate hallways. Johns Hopkins had the "Beast," and Stanford had its CART, and the Stanford Research Institute down the street featured "Shakey," who attempted to "identify and reason about large blocks" left in his path.[24]

In retrospect, Moravec says it's obvious why early robots were so thick and clumsy. Computers had seemed such geniuses, at first, because, at first, all we wanted out of them was math, something the architecture of their circuitry makes easy and the architecture of ours makes hard. But although we weren't so good at long division, we had evolved over a long time to, say, identify and reason about large blocks in our path. A computer would need to be able to carry out a thousand million instructions per second (MIPS) to match the edge, motion, and light detection capability of the human retina. To match the whole brain would require about 100 million MIPS. As late as 1999, though, an average personal computer only matched the size of an insect's nervous system—still far less than our retina alone.[25] But no need for despair; remember our old friend exponential growth. Two or three more decades of Moore's law, Moravec calculates, will be more than enough to close the gap.

What's more, scientists eventually started figuring out better ways to use the computing power they already had. Lack of speed wasn't the only reason robots tripped over blocks; they were also failing to sense the world as even modestly powered creatures could. Rodney Brooks—who found himself at Stanford working on CART at the same time as Moravec—began to ponder the problem of why, given their insect-sized brains, his robots were still not as maneuverable and adaptable as, well, insects. Instead of trying to stuff his robot brain with as much information about the world as possible, Brooks began to "eliminate any process of reasoning, or going through chains of thought." First he built Allen, and Allen begat Polly, who "was designed to be a tour guide for visitors to the MIT AI Lab" where Brooks now worked. She patrolled the corridors, trying to detect passersby so she could offer to show them the grounds. In short order, Brooks recalls, all of the grad students grew sick of her tinny voice synthesizer and would skulk past at the edge of the hall to avoid being seen. "Visitors on the other hand, we reasoned, would be excited to see a robot wandering the corridors and would most likely stop to look at it. So Polly treated anything that was in the middle of a corridor, and tall with vertical edges for the first fifty centimeters from the ground, as a potential visitor." She would tell them to shake a foot in front of her camera if they wanted a tour; if the object she had addressed was a packing crate, the lack of response would send her toddling off in search of another victim.[26]

The point was less Polly's charm than the fact that she'd been built without the accurate internal model of her surroundings that programmers had tried to install in earlier robots. Instead, she was "swimming in a sea of uncertainty." Polly in her turn begat Kismet, a robot that not only navigated but got "lonely," got "bored," got "happy." For instance, if nothing had happened to it for a while, its "boredom" drive would begin to climb, triggering it to start looking around, sort of the way you do on a dull afternoon at the desk. "At the same time, the weighting its attention system has on saturated colors will be high. So if, as it searches, Kismet happens to see some bright colors in its periphery," toward them it will go. "To an external observer, it looks like Kismet was looking for a toy and has just found one." The robot reacts to tone of voice, and although it knows only nonsense syllables, it takes its turn speaking. A visitor to the lab, a man named Richie, chatted amiably with Kismet for half an hour. "I want to show you this watch my girlfriend gave me," he said, and Kismet obediently focused on his wrist. But "before long

Kismet's attention system decided that Richie's face was more interesting than a small patch of motionless skin color. It looked back at his eyes just as Richie was done speaking, and took its turn to speak. A completely natural interaction. Kismet . . . acts like it is alive."[27] A theologian working with the AI lab contemplated the possibility of someday baptizing such an emotobot.[28]

Combining the new programming skills and the accelerating hardware, Moravec and Brooks predict ever faster growth. Moravec's lab already sells "navigation heads" to turn industrial vehicles—forklifts or floor cleaners, for instance—into robot craft. The income from those $10,000 machines will underwrite the development of home robots, beginning with vacuum cleaners that learn their way around a house, and continuing through machines to pick up clutter, take inventory, guard homes, "play games," and so on. By 2010, he predicts, the first "broadly competent universal robots" will be on the market—as big as people, perhaps, but with lizardlike brains. The next generation will be "mouselike"—able, that is, to adapt more easily, and "even be trainable." Ten years later, monkeylike brains capable of 5 million MIPS should "let a robot learn a skill by imitation, and afford a kind of conscience. Asked why there are candles on the table, a third generation robot might consult its simulation of house, owner, and self to honestly reply that it put them there because its owner likes candlelit dinners and it likes to please its owner." Fourth-generation robots, with humanlike reasoning and abilities, might take forty years to arrive, Moravec calculated in 1999; but then he added that his time scale was "probably too timid," and provided a Web address for those interested in updates.[29] Brooks contents himself with predicting "killer applications" for robots by 2010. "By the year 2020 robots will be pervasive in our lives."[30]

Meanwhile, right now, you can buy Sony's robot puppy Aibo, or Hasbro's robot doll, which Brooks helped build and which is (somewhat confusingly) called My Real Baby. Honda takes out full-page magazine ads for Asimo, a child-sized robot that can walk, climb stairs, and perform small household tasks. At the last Robodex exposition in Japan, Sony's "dream robots" jumped, danced, and kicked balls.[31] Robot airplanes have flown across the Atlantic without incident, and a robot car drove safely from Philadelphia to Los Angeles, across interstates and through traffic jams.[32] Increasingly, designers are trying to make their industrial robots autonomous; a USC professor is at work on a helicopter smart enough to spot a good landing site and then touch down, all by itself.[33] *Time* magazine gave one of its awards for the best invention of the year in 2001 to "Slugbot," a two-foot-high machine that patrols gardens using an image sensor, beaming red light, to pinpoint slugs, which emit a different infrared signal than worms or snails. Slugbot then uses a carbon fiber arm to pick up the slugs and store them in a fermentation tank, which turns the animals into electricity. It is, not to put too fine a point on it, a flesh-eating robot that can support itself.[34] Robots have even begun to add reproduction to their repertoire: in the summer of 2000, two Brandeis scientists reported in *Nature* that their robot could produce an eight-inch-long contraption of plastic bars and ball joints, and then add a microchip and a motor so the contraption could crawl off on its own. "It's a rather primitive example, but it's the first step to something that could be quite significant," said Dr. Philip Husbands, an AI specialist at the University of Sussex in England.[35] Not to worry, though, says Brooks. For a complicated robot to reproduce, it would have to be able to manufacture its own computer chips, a process that currently requires enormous and expensive factories. Therefore, one can safely assume that robot reproduction will "involve a lot of human intervention *for the next twenty-five years* [italics mine]."[36]

It has probably occurred to you that computers aren't likely to stop growing in power and speed simply because they match the human brain. Indeed, the enthusiasts say, Moore's law implies that computers will keep right on progressing, to the point where they, and the robots they inhabit, have left us behind.

The idea that machines could think, as opposed to count, has always been controversial. The great mathematician Alan Turing, who helped build the machines that broke the German submarine codes during World War II, speculated shortly after the war that a computer could successfully imitate all the functions of the human brain. Sometime around the year 2000, he guessed, computers would have a billion bytes of memory and with it be able to fool people into thinking they were human. He proposed the so-called Turing test: a human judge conducts a conversation via teletype with another human being and with a computer. If, after five minutes of chat, the judge couldn't tell which was which, the machine could be declared an "intelligent thinker." Machines now boast a billion bytes, but they still can't pass the test unless the conversation is confined to narrow themes; some experts guess computers will need to be a thousand times more powerful, which means another decade or two of development.[37]

Ray Kurzweil's timeline pretty much agrees. Though he remains an active inventor and entrepreneur, he spends more of his time these days on the big picture. He sits on the board of nanotech firms; his KurzweilAI.net is a clearinghouse for high-tech developments; and his book *The Age of Spiritual Machines* provides one of the clearest guides to this coming age. To wit, by 2009 "most routine business transactions take place between a human and a virtual personality." By 2019, "most interaction with computers is through gestures and two-way natural-language spoken communication . . . people are beginning to have relationships with automated personalities as companions, teachers, caretakers, and lovers." By 2029, "there is growing discussion about the legal rights of computers and what constitutes being human. Machines claim to be conscious and these claims are largely accepted." And by 2099? Well, by 2099 "there is no longer any clear distinction between humans and computers."[38] A penny's worth of computing power then will be a billion times as powerful as all the human brains now on the planet.[39]

Such claims rely on extrapolations deep into the future (on the notion that as they get wiser, the machines begin to "self-bootstrap," improving themselves "in ways that limited human minds can't even start to understand").[40] And scientists remain—though in shrinking numbers—who insist that consciousness can't be approximated by a machine, that there is something different about minds that mere increases in computing power can't approach. As we shall see in the last chapter, I, too, think there is at least one way in which the human mind is unique.

But we need to take seriously the possibility that if we allow computers, and hence robots, to become steadily more powerful, they really will roar right by us. Consider the testimony of Gary Kasparov, the greatest chess player who has ever lived, and also the only human who has ever faced off against a machine expressly built to exceed his powers. (It's true that chess is "only" a test of computational ability. But it's also true, as Csikszentmihalyi discovered, that many people have made it a center of their emotional lives.) When IBM's Deep Blue beat Kasparov in the first game of their 1996 match, it felt like a *force,* not a machine. As Charles Krauthammer wrote at the time, late in the game Deep Blue's king was

> under savage attack by Kasparov. Any human player under such assault by a world champion would be staring at his own king trying to figure out how to get away. Instead, Blue ignored the threat and quite nonchalantly went hunting for lowly pawns at the other end of the board. . . . It was as if, at Gettysburg, General Meade had sent his soldiers out for a bit of apple picking moments before Pickett's charge because he had calculated that they could get back to their positions with half a second to spare.

Indeed, the computer had done precisely that: it knew it had nothing to fear because, looking over 200 million possible positions per second, it "determined with absolute certainty that it could return from its pawn-picking expedition and destroy Kasparov exactly one move before Kasparov could destroy it. Which it did." Only a computer could do that; "no human can achieve absolute certainty because no human can be sure to have seen everything."[41] Which may have been what Kasparov had in mind after the sixth and final game when he said, "I was not in the mood of playing at all. I am a human being. When I see something that is well beyond my understanding, I'm afraid."[42]

At about the same Eisenhower-era moment when those first slide-rule-toting computer jocks were building their earliest robot carts, the physicist Richard Feynman planted another seed that is also bearing fruit just about now. In 1959, he gave a little talk to the American Physical Society with the title "There's Plenty of Room at the Bottom." He began by asking, "Why cannot we write the entire 24 volumes of the *Encyclopaedia Britannica* on the head of a pin?" and went on to demonstrate how you could, indeed, reduce every dot on the printed pages to the requisite size, thirty-two atoms across. Feynman proved that you could put all the books in all the world's great libraries "into a mass of metal that was tinier than a pin head," simply by transcribing the words into dots and dashes, each about five atoms long, and then writing them not only on the pin head but on the layers of metal beneath. "It turns out," he said, "that all of the information that man has carefully accumulated in all the books of the world can be written in this form in a cube of material 1/200th of an inch wide—which is the barest piece of dust that can be made out by the human eye. So there is *plenty* of room at the bottom. Don't tell me about microfilm."

Feynman was outlining what we now call nanotechnology. For all its exotic sound, it is, as he pointed out, simply the next step in miniaturization, the next step down the scale from microtechnology. A micrometer is a millionth of a meter, and a nanometer is a billionth of a meter. But what a step that is. Once you're in that realm, he pointed out, you needn't confine yourself to parlor tricks of miniaturization. Once you had a mechanism for moving atoms around one by one to, say, print encyclopedias, *you'd be able to make any material you wanted.* "Give the orders and the physicist synthesizes it. How? Put the atoms down where the chemist says, and you make the substance."[43] If you wanted a water molecule, you'd move some hydrogens into place around an oxygen. Of course, water is already easy to come by. But the same basic technique would give you diamonds. Or, for that matter, carrots. They're all made up of stuff. Seen that way, in the words of the agriculture activist Pat Mooney, "nanotechnology is to inanimate matter what biotech is to animate matter."[44] Seen that way, anything becomes possible.

But it took people a while to see it that way; we weren't quite near enough the knee of the curve. For some decades, explains B. C. Crandall, the president of Mimetic Engineering, chemists concentrated on molecular science, or seeing the world from the bottom up; hence nylon. Physicists and electrical engineers, meanwhile, built ever smaller machines from the top down; hence the microchip. It's only in recent years that those disciplines have begun to converge, and they are doing so faster all the time. While "computer-chip designers build a fairly small number of complex machines as minutely as possible, chemists build relatively simple but atomically precise machines by the billions. Nanotechnology rises out of this confluence and aims at building complex, atomically precise machines by the trillions."[45]

The year 1981 marked an early watershed for the technology, when a pair of IBM scientists invented the scanning tunneling microscope, which offered the first direct images of individual atoms, and K. Eric Drexler, then a researcher at MIT, published the first of his papers arguing that the natural mechanisms of protein synthesis proved we should be able to build molecule-sized machines. A year later he introduced the idea to the general public in an article in *Smithsonian* magazine, and by 1988 he was teaching the first college course on nanotech, at Stanford.[46] The subject finally broke in upon the public consciousness in 1990, when IBM spelled out its company logo by moving thirty-five xenon atoms into the appropriate spots on a nickel crystal.[47]

So far, about three hundred companies here and overseas are hard at work on nano-engineered products: Nanogram, NanoOpto, Nanophase, Nanosphere, Technanogy.[48] The items they're making are fairly mundane: sun lotions with invisibly small zinc oxide particles, coatings that make eyeglass lenses more sun resistant. Next in line will be super-tough nanofiber-reinforced plastics, stronger and lighter than steel. For the aerospace industry, said one executive, "its importance would be almost a bigger step than going from propellers to jets." Diabetics will be able to draw blood and inject insulin through teeny "nanostraws" that cause no pain.[49]

All these things are sensible, incremental; they don't threaten to upset the world any more than the microscale revolution of the last fifty years upset it (a lot, in other words, but perhaps not unmanageably). They are the equivalent of somatic genetic engineering, the kind that is used on existing human beings to cure existing problems. They are Alexander Graham Bell inventions—powerful, but within our general paradigm. The gravity of our civilization could contain them.

But the holy grail of the nanotechnologists, the thing that makes the futurists' eyes light up and twirl around, the mechanical equivalent of germline engineering, is the so-called assembler. An assembler would be a machine roughly the size of a strand of DNA, able to move individual atoms around and put them precisely where you wanted them. With a programmable assembler, one that could move billions of atoms to the right place at the right time so they would stick together in molecular structures according to the laws of chemistry—well then, you really *would* able to build anything. Probably the first thing you'd build would be more assemblers. The technology would go from being small but inert to being small but almost . . . alive, a self-replicating machine far more powerful than anything we've ever let loose. That self-replication creates unique practical dangers, as we shall see, but it also raises even more profound questions simply because of its all-encompassing power. Here's how K. Eric Drexler imagines it might happen: Working at a million atoms per second, an assembler would be able to copy itself in a thousand seconds—about fifteen minutes, or the time a bacterium takes to replicate. And then each of those assemblers could build a copy of itself, and so on and so on and so on, till eventually you would have something big enough to be of use to humans. In ten hours, for instance, the assemblers could replicate 68 billion times. You begin to see the power.

In essence, you would have a machine reproducing material at almost no cost, the way computers reproduce information. You could, theoretically, take such a universal assembling machine and toss some grass clippings in one end. The replicators would be able to seize the carbon in that grass, breaking down certain chemical bonds and constructing others, all according to a plan. In the words of the science writer Ed Regis, "after a while you'd open the box and out would roll a wad of fresh beef."[50] Ditto with paper, or filing cabinets, or potatoes. Kurzweil puts it succinctly: "They'll solve humanity's material needs."[51]

There are some engineers who caution that we don't know quite how assemblers will be able to manage energy supplies, read their programming instructions, and so on.[52] Researchers talk about the "fat fingers" problem—that "picking up and moving atoms involves building instruments at least as big as the things they are moving." Others caution that so far atomic manipulation has worked best in a vacuum at temperatures approaching absolute zero (which is to say, 459 degrees below 0 degrees Fahrenheit).[53] Drexler concedes that there will be problems with vibration and heat in the close quarters of atomic manipulation. But, he stresses, "I have yet to encounter a major technical criticism of the core concepts of nanotechnology that does not evaporate once it is examined."[54] In a decade of trying, no one has cited any physical laws this work would violate. It is, after all, the way that food is made at present: potato plants and grass break down soil and water and use solar energy to rearrange them; cows conspire to convert the grass into burger. The basic cost of nano-assembled stuff (minus, of course, the licensing agreements you'd pay to the various inventors) would be the cost of the raw materials. And dirt is notoriously cheap.

What could we conceivably do with it? Well, almost anything. Medical "nanobots" might cruise our bloodstreams, attacking pathogens within our bodies and building new cells, even new organs.

There might be an end to agriculture, and to air and water pollution, and to the use of fossil fuels.

And, of course, we would see "the provision of new and theoretically limitless consumer products."[55] At first, manufacturers will make existing products in new ways; if you make your living producing cotton bath towels, say the editors of *Nanotechnology*, you need to start "getting a handle on just what a bath towel physically looks like on a molecular level, figuring out the proportions of carbon, oxygen, etc., and writing the software for your full line of towels with all their various colors, weaves, and patterns." Then, when your assembler arrives, you'll stop paying farmers for cotton, start paying whoever holds the patent on the machine, and hook it up to a pile of carbon black. In will go atoms and out will come terrycloth.[56]

But who needs a bath towel when nanotechnology might also allow "people-scrubbers that run off your body heat and constantly scrub dirt and odors from your body, so you never need to take a shower"? Or, to give just a small sample of the other delights enthusiasts have imagined, full-wall video screens, self-adjusting contour chairs that change shape to match the person sitting in them, "board games with billions of moving parts," "reversible shrink-wrap bags that seal and unseal on command," "active food that squirms around as you eat it but becomes inert once swallowed," and spray-on pants.[57] Or imagine antiwrinkle nanomachines designed to "improve the skin's elasticity" by making sure it keeps producing collagen (so much for Botox).[58] A nanomachine might insert itself into the melanocytes of hair follicles, "where it benignly but persistently regulates the production or delivery of the pigment melanin. . . . As a result, a brunette could bleach her hair once and never again have to worry about dark roots growing out." Or she could change the color of her skin. ("There can be little doubt . . . as to the market potential" of that invention.) Perhaps a permanent breath freshener, made of "a solution teeming with nanomachines that cling tenaciously to every bit of exposed tooth enamel and link up to form a complete, invisible surface film around each tooth."[59] (Dentists, for some reason, are particularly fascinated by nanotechnology. In his essay "Dental Care in the Nanofuture," Edward M. Reifman imagines Dr. Harvey Smile-Maker "peering into the oral cavity of his new patient, Mr. John Garbage-Mouth," who refuses to floss, rots his teeth with chocolate, and just lost his girlfriend because of his "horrific mouth odor." Luckily for him, Dr. Smile-Maker has his new nano-equipment on hand, and "before you can say 'dental floss' he grows his patient a new tooth."[60])

"What we are talking about here is replacing our current physical technology for manufacturing with a whole new set of materials and devices that have far superior computational capacity and power," says Drexler. "The closest thing we have seen to this type of transformation was the industrial revolution, and that comparison somehow doesn't seem adequate. It's very hard to imagine a future that will involve that much change—it's exhausting, so the common reaction is to say: 'This is indigestible, therefore I am not going to digest it!' "[61]

We will try to digest it in a little while, but first it makes sense to ask how fast it's really coming: whether it's an issue for our digestive tracts, or those of our grandkids.

On the one hand, all the first-order applications of nanoscale technology, things like the sunscreen with the little particles of zinc oxide, are just exploding. You can get stain-resistant "Nano-care" khaki pants from Eddie Bauer, for instance, that contain nano-sized surface fibers.[62] Computer chip makers have developed nanoscale transistors and prototype nanotube circuits. "There's geometric progress in this area," said one Harvard professor. "A year ago I would sort of wink at people when I said 'nanoelectronics.' Now, I actually believe in my heart we'll be able to do it."[63] The early developments are shaking loose private money; "there has been a sharp upsurge in the number of venture capitalists at science meetings," says the head of the nanotechnology division at Mitre Corporation,[64] and the National Science Foundation recently predicted that nanotech could be a $1 trillion market by 2015.[65] IBM reportedly devotes half its long-term research budget to nanoscale projects.[66] High school students now build scanning tunneling microscopes as science fair projects.[67]

On the other hand, we've not yet crossed the Rubicon here either, and it's perhaps a little further away than with human genetic manipulation. While there's plenty of targeted niche research under way, fewer people are doing the fundamental R&D necessary to, say, build a universal assembler. "We're at a scientific halfway point," in the words of the journalist Nicholas Thompson. "We know the technology is plausible, and even likely, but we haven't advanced enough for private industry to commit more money. Scientists still have to trudge pretty deep into the jungle, and they are likely to frequently step in quicksand." If it's going to happen, he reports, "the government has to be a major player. It just won't happen in industry."[68]

Government has begun to respond. Bill Clinton, in January 2000, appeared at Caltech, standing before a backdrop that showed the Western Hemisphere painted in individual gold atoms, and announced he would support "a major new nanotechnology initiative worth $500 million."[69] A few weeks earlier, his science adviser, Neal Lane, had predicted that nanotech would have as big an impact as "information technology, or cellular, genetic, and molecular biology."[70] (Not long after leaving office, Clinton ventured out to Aspen, Colorado, to give a talk at a symposium called Brainstorm 2001. He commanded his audience to read *The Age of Spiritual Machines,* calling Kurzweil's vision of, among other things, a completely nanotechnological world a "compelling view of the future."[71]) Clinton wasn't by any means the only politician to be drawn in; by September 2002, senators in a hearing were examining nanofabric trousers and weighing a bill to establish a "National Nanotechnology Research Project."[72] And America is far from alone. In 2001 the Chinese opened a Center for Nanoscience at the Chinese Academy of Sciences; more than a dozen universities across the country are already collaborating on projects that have produced nanofibers at sixty times the old rate and built the world's smallest nanotube.[73] The Japanese, meanwhile, looking to repair their fortunes, have launched what *Nanotechnology* magazine called "the best funded research program on earth," and the French and Germans have created what they call Nano-valley in the upper Rhine.[74] Harvard and MIT and all the other obvious places have made big investments in nanotech research—but so, for instance, has the University of Texas at Dallas, backed by a state legislature that "wants to make North Texas into the nanotech equivalent of Silicon Valley."[75]

North Texas is already home to the Zyvex Corporation, which bids to become the IBM, the Ford, the Boeing of this new technology. No nano-sunscreen for Zyvex, whose board of directors includes not only Kurzweil but also James L. Halperin, author of a paean to cryogenics called *The First Immortal.* James Von Ehr, a Texan who'd made a good-sized fortune in desktop publishing software, assembled this group of true believers after he heard Drexler give an after-dinner speech. Zyvex, the world's first nanotech startup, has one goal in mind: to build an assembler "which others can buy and set to work to make useful things for the markets they are already familiar with. We don't expect to conquer all known markets and make all possible products by ourselves. The assembler will be enough for starters."[76] As indeed it would. The earliest models would be bulky and not capable of "building a diamond spaceship out of potato peelings," producing instead small quantities of incredibly lightweight and strong nanotubes.[77] But they would more or less prove the concept. Alexander Graham Bell didn't exactly invent the cell phone, but if you'd been in his lab that afternoon you might have felt the world shake.

So when will we start to see the earliest such devices? Zyvex says five to ten years. *Nanotechnology* magazine says eight to fifteen.[78] When *Wired* magazine polled a group of the top scientists at work on nanotechnology, they found guesses averaging between 2010 and 2015.[79] Investors believe: a reporter covering a recent nanotech conference at Harvard Business School said, "The feeling among this small group of people that track the industry is that all the hype about the potential is probably true—it really is the next big thing."[80] And the believers aren't just boosters touting their start-up stock. In 2001, Pat Mooney published a remarkable comparison of biotech in 1987 and nanotech in 2001. Just fifteen years ago, he pointed out, many scientists still wondered if genetic engineering of plants and animals wouldn't "run foul of the infinite complexity of nature," just as some researchers believe "manipulating the Table of Elements" might violate "still unknown natural laws." But by now, Mooney points out, hundreds of millions of acres are sown with genetically modified crops. Atoms "are the next logical declension from genes."[81] And in this case Fortune 500 companies are getting in on the ground floor. One more thing: if such devices start appearing in 2020 or 2030 instead of 2015, it won't make much difference in the long run.

Exactly what happens when almost certainly makes less difference than the general trend, especially because all these technologies are racing toward some murky convergence. As long ago as 1984, Thomas Pynchon, a novelist with sensitive antennae, wrote in the *New York Times Book Review* that "if our world survives, the next great challenge to watch out for will come—you heard it here first—when the curves of research and development in artificial intelligence, molecular biology, and robotics all converge. Oboy."[82]

"Oboy" in this case might mean, for instance, the construction of "nanobots," swarms of robots the size of microbes able, among other things, to wander the human bloodstream keeping us forever healthy.[83] And maybe "mind uploading," the ability to "scan the human brain and re-create its design electronically," perhaps by using those same nanobots to "capture the locations, interconnections, and contents of all the nerve cell bodies, axons, dendrites, pre-synaptic vesicles, neurotransmitter concentrations and other relevant neural components" and sending the information to a hard drive via wireless transmission.[84] When they're done, perhaps "you" will be able to reside in a computer—or, at least, a pretty good backup will.

No one knows whether "mind-uploading" is really possible—whether you can actually capture memories and feelings in such a fashion. But clearly you can work the trick the other way around, implanting hardware in your brain to hook you up to electronics. So far, engineers have done some very useful things: cochlear implants, for instance, that restore at least some hearing to the deaf by transmitting electrical impulses directly to the brain. But "these medical attempts to repair the body might just be the opening wedge in an expanding industry dedicated to extending the mental and physical functions of human beings through electronics." The body modification movement, writes the science journalist Steve Edwards, "could get a tremendous boost from implanting math co-processors, hard drives, miniature video-cameras, infrared vision, cell-phones. The mind boggles at the possibilities."[85]

If the mind boggles so much as to doubt, however, it's worth noting that scientists have already "intercepted the neural transmission of a cat's eye from *behind* its optic nerve," meaning the cat had already done some processing of the image. Using "linear decoding technology," they translated these brain waves into "grainy but identifiable images. Monitors displayed what the cat was seeing."[86] In 2001, New York researchers fitted live rats with three wires in their brains. A pulse of current along one made it feel as if its right whiskers had been touched; another tickled the left whiskers; and the third was connected to the brain's "pleasure center," the medial forebrain bundle. Then, with a joystick, the scientists were able to make the rats scramble any which way they wanted; though the animals had never before been outdoors, they were quickly climbing trees, scurrying along branches, and so forth. They had, in other words, solved the navigational difficulties faced by robotics researchers by turning their rats *into* robots.[87] Researchers at the University of Genoa did the same thing in reverse, creating a mechanical robot fish whose movements are controlled by the brains of an eel.[88]

Ian Pearson, employed by British Telecom as its official "futurologist" and charged with "tracking technology developments across the whole field of information technology" to "develop future scenarios for his employer," offers a typical forecast. "Homo Cyberneticus" will soon emerge with a "full duplex link between man and machine." This creature will in turn merge with "Homo Optimus," the genetically engineered "elite race of people who are smart, agile, and disease-resistant." Together they will form "Homo Hybridus," which will have no trouble displacing "Homo Ludditus."[89] You might be forgiven for imagining it all as a race. Will humans be turned into robots before robots can be turned into humans? Which of them will get to use the nanomagic? And you might be forgiven for asking a deeper question: long before the race is over, will either "robot" or "human" have any meaning?

But before we get to questions of meaning, which are really questions about what happens if this technology works the way it's supposed to, it's probably worth pondering another matter: what if these technologies go wrong?

Bill Joy could be Hans Moravec could be Ray Kurzweil could be Rodney Brooks. He, too, grew up in the thrall of technology: "Thursday nights my parents went bowling, and we kids stayed home alone. It was the night of Gene Roddenberry's original *Star Trek*, and the program made a big impression on me." He, too, spent his college years writing code. And he, too, was recognized as a true prodigy early on, except that he made even more money, had even more impact. After inventing the Berkeley version of Unix, which became the backbone of the Internet, he went off to Sun Microsystems, where, as chief scientist, he led the teams that wrote Java and Jini.[90] He is, in other words, the real deal—"a radiant figure in many ways outshining any other in the industry," said one high-tech veteran. "His intellectual attainments far exceed those on the short lists of recently newsworthy brains such as Bill Gates, Nathan Myhrvold, David Boies, or Warren Buffett."[91]

And *Wired* magazine is the real thing too, the closest thing to a bible Silicon Valley has ever had, for it not only keeps tabs on everything new, it also keeps the faith. Each issue hypes the technofuture, promising an "endless boom" of prosperity as the new gadgets work their way into the fabric of our lives.

So it brought people up short when the April 2000 issue arrived with a cover article by Bill Joy that, in effect, said that society should relinquish human genetic research, nanotechnology, and advanced robotics. It's as if *Car and Driver* landed in the mailbox with a lead story from Bill Ford, Jr., extolling the virtues of walking. It's as if *Mother Earth News* assigned Ben and Jerry to do an exposé on the link between granola and cancer.

Joy began by stating his techie credentials: "I trust it is clear that I am not a Luddite. I have always, rather, had a strong belief in the value of the scientific search for truth and in the ability of great engineering to bring material progress." And he applauded the aims of the researchers at work in the fields he called GNR (genetics, nanotechnology, and robotics). In fact, he said, his early reading about nanotechnology had left him "with a sense of calm—it showed us that incredible progress was possible, and indeed perhaps inevitable. If nanotechnology was our future, then I didn't feel pressed to solve so many problems in the present. I might as well enjoy life in the here and now. It didn't make sense, given this vision, to stay up all night, all the time" designing better computer languages.[92]

In recent months, however, Joy said he'd been jolted by colleagues telling him that the timeline for these new sciences was much more compressed than he'd thought. The human genome was very nearly sequenced, new developments meant that "nanoscale molecular electronics was now possible," Kurzweil convinced him over drinks that intelligent robots were "a realistic and imminent scenario" thanks to Moore's law (which had, of course, powered Joy's own achievements in computer science.)[93] He began to think again, and this time he wasn't so calm; in fact, he became convinced that these technologies actually presented a clear and present danger.

The danger arose in part from their pure power—"I mean, the people doing genetic engineering are doing really good stuff. It's just that I don't believe they've dealt with the confluence of their field and Moore's law. . . . Within 20 or 30 years, probably sooner, if we give everybody personal computers that are a million times as powerful, they'll have the ability to manufacture whatever they can design. We'll have an untenable situation because people can design new diseases and build them on their own computers."[94] But the deeper problem, Joy felt, was that the three technologies had a common flaw: they could self-replicate. Tailored germs could spread; robots could build more of themselves; and as for nanotechnology, the ultimate promise of the field depends, as we've seen, on the replication of the tiny assemblers. "It is most of all the destructive self-replication of GNR that should give us pause," Joy wrote.[95] Their multiplying ways made even the nuclear and chemical weapons of the twentieth century seem tame by comparison. Bad as atom bombs were, you had to build them one at a time.

It wasn't the first time someone had raised the issue. Joy, in fact, sent readers back to read one of the final chapters of K. Eric Drexler's original opus on nanotech, *Engines of Creation.* The passage concerned the nasty possibility that nanobots—which, after all, would be designed to eat everyday materials in order to construct their wonders—might get too hungry and "reduce the biosphere to dust in a matter of days." The small group of Drexler devotees had spent much brainpower over the past decade debating this so-called gray goo problem, which soon expanded to include the gray plankton, gray dust, and gray lichens dilemmas, as well as the ominous-sounding "malicious ecophagy." Their papers swelled with equations and jargon: "The highest near-term risk could come from relatively simple single-behavior replibots whose niche is a high-energy substrate of uniform composition which affords a rapid vector for the dispersal of replicators," wrote Robert Freitas, a Zyvex research scientist, in a typical paper. A badly or malicously programmed bot strain might decide not just to clean up a tire dump but to eat the rubber in tires and asphalt tar binder. Everywhere. Or all the cotton. Or polyester, or insulation on electrical wiring, or "paper money." In a worst-case scenario, the replicators might not stop with tires and dollar bills, but instead

> convert the entire surface biosphere (the ecology of all living things on the surface of the Earth) into alternative or artificial materials of some type—especially, materials like themselves,

> e.g. more self-replicating nanorobots. . . . Ecophagic nanorobots would regard living things as environmental carbon accumulators, and biomass as a valuable ore to be mined for carbon and energy. Of course, biosystems from which all carbon has been extracted can no longer be alive but would instead become lifeless chemical sludge."[96]

Of course. Or, as one Montreal newspaper columnist imagined, a leak from a Canadian nanotech lab lets several tiny nanobots escape into the downtown smog. "Within hours, all of southern Ontario is encased in a swirling, exploding mass of metal."[97] Bad, eh?

The nano-enthusiasts also occupied themselves considering whether such "gray goo" might be effectively countered by "blue goo," policebots that would form a nanotechnological immune system. But perhaps some of the blue goo would turn gray? And wouldn't even the blue goo use too much energy?[98] "Goodbots" might enjoy technical advantages over "badbots," argued some: for instance, "while badbots must simultaneously replicate and defend themselves against attack, goodbots may concentrate exclusively on attacking badbots and thus enjoy lower operational overhead." The most ambitious calculations indicated that a sneak gray goo attack would be detected, because in the course of eating up the biosphere the nanobots might raise the global temperature by several degrees, which is perhaps not the most comforting scenario one can imagine.[99]

Faced with all this, Joy bravely and eloquently called for "relinquishment" of these technologies. "Yes, I know, knowledge is good, as is the search for new truths. . . . We have, as a bedrock value in our society, long agreed on the value of open access to information. In recent times we have come to revere scientific knowledge. But despite the strong historical precedents, if open access to and unlimited development of knowledge henceforth puts us all in clear danger of extinction, then common sense demands we reexamine even these basic, long-held beliefs."[100]

His call was met mostly with indignation or resignation from his peers in Silicon Valley, many of whom are libertarian in their instincts, viscerally opposed to the kind of government regulation that any relinquishment scheme would require. But to me his essay seems one of the great Paul Revere moments of our time, a full-throated and unhesitating alarum that should scare the hell out of us. We know enough about technology and unforeseen consequences to sense its truth: in July 2002, a team of scientists working only from diagrams in a book managed to create an infectious poliovirus. Smallpox, they say reassuringly, would be somewhat harder to fabricate.[101]

We would be insane to take risks like this. As with genetic engineering, though, even these are not the greatest perils. Nanotech and robotics not only threaten our survival as "environmental carbon accumulators" but, in their most advanced forms, resemble germline engineering: they are simply too strong to mesh with the human condition as we have known it. They threaten—indeed, they promise—to destroy the meaning of our lives.

It has perhaps crossed even your dull old-fashioned human mind that there is something of an irony at work here. Even as the genetic engineers work busily to upgrade us, adding IQ and memory, the robotics engineers are hard at work making sure we'll be surpassed, and the nanotechnologists to make sure all our wants will be satisfied by pushing buttons. What, in other words, are we being enhanced *for*?

Since the dawn of robotics, people have feared they'd be put out of work by automatons. Some have, of course—spot welders on automobile assembly lines, for example, were easily replaced by machines able to do the same job over and over with precision. In the past decade, more and more jobs have shifted over: telephone operators replaced by voice-mail systems, and bank tellers rendered obsolete by ATMs. The trend accelerates: already fast-food restaurants in Japan have tested "server robots," and "simple, self-guided robots with TV cameras able to recognize intruders are finding work as security guards." When two airplanes crashed over Germany in the summer of 2002, the verdict was swift: the pilots had erred by following the instructions of an air traffic controller rather than the commands of their automated collision avoidance system. "Someday, no airline will dare put a person at the controls for fear of lawsuits."[102] If, with Moore's law on their side, robots become ever better at handling materials, ever better at processing information, and ever ever cheaper, then they "may displace human labor so broadly that the average workday would have to plummet to practically zero to keep everyone employed."[103] This may seem a distant prospect—but if we're actually in the knee of the curve, remember, time collapses.

Whenever it happens, such a future raises certain practical problems, of course—like, what do you for cash? The roboticists and nanotechies have done their best to think the problem through. Ian Pearson, the British Telecom forecaster, anticipates that since enterprises run entirely by machine will easily outcompete human workers, "people will have fewer and fewer attributes to sell." The only consolation is that "production and output could greatly increase . . . so we could all have a better quality of life without having to do work."[104] Perhaps, suggests Moravec, an expanded Social Security system would levy high corporate taxes on the roboticized industries; then, "by gradually lowering the retirement age, most of the population would eventually be supported. The money could be distributed under other names," he notes, "but calling it a pension is meaningful symbolism. Social Security payments begun at birth would subsidize a long, comfortable retirement for the entire original-model human race." In such a scenario, "the primary job of humanity in [this] century will be protecting its retirement benefits by ensuring continued cooperation from the robot industries."[105]

That's a powerful image, a whole world of people "in retirement" from the moment they're born. It raises the stakes; the problems the technologies pose turn out to be not only practical but essential. Even if such a scheme worked economically, how would it *feel*? Work is one of the things that orders our lives. If it is sheer drudgery, it may dull and shorten our lives; if there is too much of it, we may feel as if there are other experiences we're missing. But for the most part, the chance to develop skills and to apply them, to see our sweat manifested not only in a paycheck but in a harvest, a house, a book, a classroom full of growing children—that is among the strongest day-in, day-out meanings of our lives. Already we've taken our bodies pretty far out of the equation; most of us in the Western world no longer earn our bread from the sweat of our brows, the bend of our backs. But even in the less demanding jobs we do instead, much of the strange connection between effort and joy and pride and reward still remains. Each of us knows the difference between doing good work and going through the motions, and each of us knows how those differences make us feel when the day is through. The most liberal of us understood there was something debilitating about welfare—knew what its critics were talking about when, instead of moaning about their tax dollars, they talked about how the dole robs people of a certain kind of dignity. The man who dies six months after retiring is a cliché; so is the retiree who can't quite make golf, or puttering around, or watching TV fill the hole in his life.

Having nothing to do is one kind of hell. Never mind that you still have stuff, more stuff than ever. In a way, that makes it worse. You remember the cargo cultists, those South Sea islanders who during the Second World War were briefly invaded by Western civilization and all its attendant stuff? Who thereafter devoted themselves to praying that the planes would return, with their crates of canned food? Who drew jets in the beach sand to lure back the GIs? "To have without doing corrodes the soul," writes Erazim Kohák, a perceptive philosopher. "It is precisely in investing life, love, and labor that we constitute the world as personal, as the place of intimate dwelling."[106] Wendell Berry, the Kentucky farmer and writer, once titled an essay collection *What Are People For?* He suggested the answer elsewhere, in a truly lovely short story, "The Boundary," that describes an old man and his wife living out their last years on their farm. His son has taken over the fields, but the old man plants a garden yet, and daily carries home to his wife in the kitchen half a bushel of peas or an armload of squash. "The garden pleased him.

After even so many years, he still needed to be bringing something to her."[107]

Against all this, the idea of, say, a nanotech magic box seems crude, gross. What would it mean to "give" anything to anyone (including yourself) if all you had to do to get it was press a button? Everyone on earth wins the lottery every day—and then what? In one of his essays, Berry writes, "My wish is simply to live my life as fully as I can. And in our time this means that we must save ourselves from the products that we are asked to buy in order, ultimately, to replace ourselves."[108] One nanotech enthusiast, writing in *Popular Mechanics,* declares that its power will "transform the most desolate village into a Garden of Eden, with widescreen TVs and cappuccino machines for all."[109] I know the kind of villages he has in mind because I've spent time in them. They *are* poor, and they *are* isolated, but many of them nonetheless contain an order and a life that does not require cappuccino machines appearing from the ether to make them whole, and that could not sustain endless unearned largesse without disintegrating. "A potato is disassembled atom-by-atom and the information is recorded in an outrageously powerful nanocomputer data bank," the editors at *Nanozine.com* explain in an essay titled "Nanotechnological Pursuit of Happiness." "One can broadcast this information to the other side of the Earth or to a moon of Jupiter. Then, with the right feed stocks (carbon, oxygen, hydrogen, etc.) and a few trillion nanoassemblers, one can reconstruct an exact copy of a potato."[110] I don't grow potatoes—perhaps it's horrible work that should be replaced by magic. But I have a neighbor who runs Golden Russet Farms; I buy his spuds at the co-op, and I sense that they mean something more than their atomic profile.

I *do* write books. I know the years of thinking and interviewing and gathering and outlining, and then the months of writing, and then the paring and shaping. It is work; you're tired at the end of the day. But it's good work, or can be. When I read the prophets of our technofuture explain that soon computers will be ever so much more powerful, and that for these cyberbeings "writing a book will take a minute or two," it's not the lack of a paycheck that troubles me most.[111]

How would we spend our eternal retirement? What would compensate us for these kind of losses? Marvin Minsky, a pioneer of artificial intelligence almost from its conception, once paraphrased the science fiction writer Isaac Asimov: "One thing people seem to like especially is being able to order other people around—this may be the greatest social use of robots. While there may never be enough people to serve this need, there could be enough robots. Everyone could order them around whenever this urge-to-power comes on."[112] A future as plantation overseers! Perhaps if they displeased us we could whip them; that might fulfill some other urge.

Usually, though, we're told that the tradeoff is more leisure. "Freed from work," writes Berry, "men will presumably take to more 'worthy' pursuits such as 'culture.' Noting that there have always been some people who, when they had the leisure, studied literature and painting and music, the prophets of the technological paradise have always assured us that once we have turned all our work over to machines we will become a nation of artists or, at worst, a nation of art critics."[113] Indeed, that is the standard pitch: freed from drudgery, we will have the time to devote to more "fulfilling" endeavors. But even if that's true—and so far, machinery has mostly freed us to watch more *Seinfeld*—what if those endeavors turn out to have been just as degraded as work? Why would an age that thinks of work as something to be dispensed with feel differently about, say, learning? Once we have the neural implants that researchers have already begun to perfect in rats and eels, we'll be veritable learning machines. "Unlike nature," writes Kurzweil, "we won't leave out a quick knowledge downloading port in the electronic version of our synapses." With that improvement we'll be able to overcome "the slow process of human communication, of human teaching and learning," and instead "be able to read a book in a few seconds."[114] Or if that's too slow, by the time we can implant 100,000 electrodes per square inch of scalp, there'll be "no need to read a book—the computer just squirts its contents into your head."[115]

Perhaps you'll while away your retirement contemplating "art." Maybe a show like "Artbots," held at the Pratt Institute in New York in the spring of 2002. "The show filled a warren of rooms with the whirr and whine of tiny electric motors. On the floor of one room, three robots made of Lego bricks topped with plastic dolls' heads pulled Japanese ink brushes across a scroll of paper, producing swirls of thick black strokes. 'We're no longer the artists—we're the attendants,'" explained the programmer, Eva Sutton.[116] Or maybe you'd prefer sports. Viewers in twenty countries already tune in weekly to watch *Robot Wars* and *BattleBots,* which pit homemade robots like Anthrax (a "six-wheeled heavyweight with a four-ton hydraulic jaw") against Cannibal ("a cross between a kitchen hinge and a pot of salt.") Comedy Central, which airs the American version, reports that it trails only wrestling in the cable sports ratings.[117]

But what about those who still felt some actual need to participate in life themselves? Who still possessed the atavistic urge for the kind of "flow experiences" and "focal practices" that, as we saw in chapter 1, give joy to human life as we now know it? Well, they would have lots of free time to pursue a hobby. Maybe two hobbies! Tom McKendree, in a small essay titled "Nanotech Hobbies," gives us some sense of what to expect. He begins with model railroading. Since, as he points out, a rotary engine as small as 30 nanometers in length could be constructed, one could build "a model railroad where the rails are 50 nanometers apart," giving you a scale ratio of over a billion to one. Now, since a single car would be near a light wave in length, "a train would be much too small to see directly. On the other hand, one could build a working replica—small but visible—of the entire U.S. railroad network."

Your home workshop could be stocked with tools whose "handles could change their shape to fit the hands perfectly and massage away muscle pains," and you would be able to incorporate nanotechnology in your crafts, "including, say, a video game in an ornately carved wooden board, or making a stool that was self-repairing."

For needlepointers, who like "the pleasant calming effect" that stitching produces in their souls, "nanotechnology could help by removing some of the petty annoyances—self-threading needles for example. A piece of cloth could have a pattern printed directly on it that faded away at each stitch, or sounded a gentle alarm if one mis-stitched a thread." Gardeners who didn't just want to press "potato" on their machine might use swarms of nanobots in place of pesticides, "or biological nanotechnology could be used to create new, hardy strains of plants that look and smell exactly like the original but are designed to thrive in utterly different climates. Thus, one could grow what looked like orchids next to a mountain cabin, and what looked like a healthy blue fir on a sunny beach. . . . One could even order personalized cuttings, so that each petal of a rose bore the cameo of someone in the family."

And if you want that old adrenaline rush? "Molecular engineering could provide super bungee cords of varying thickness and much stronger material that would allow significantly higher jumps." Not only that, but "hand to hand combat using medieval weapons could provide a thrill" once we'd mastered nanomedicine to the point where any damage would be easily reparable. "This activity could be quite realistic," McKendree purrs, "down to producing wounds that today would kill a person."[118] Or, as Robert Freitas points out, you could simply get a robot to play for you, looking through his eyes with "telepresence." "This could be an especially interesting prospect for highly dangerous activities you might not otherwise have the nerve to try—teleoperated boxing, racecar driving, parachuting, or mountain climbing. Telesports would let people feel reckless without risking personal harm. . . . Telerobots could even revive the dueling tradition!"[119]

Or maybe we should just shoot ourselves now, while the bullets still actually work.

Look—I don't know for sure what this world would feel like. What it would mean if we were "seamlessly articulated with intelligent machines" or if "the realm of the born and the realm of the made . . . become one."[120] I can imagine what a cat feels like stretching in the late afternoon sun, but I can't quite channel what it would be like to inhabit "a warm, energized, super-sensual morphing device of graceful complexity and beauty."[121] It's a thought experiment almost beyond our powers. What *savor* would the world hold, what crunch? It's like imagining you're a car, trying to sense the asphalt against your rubber, the fluid coursing through your brake lines. Try as I might, these seem to me deadening, muffling technologies, without even the naïve promise of liberation that accompanies some of the talk of genetic engineering. They would cut us off even more fully than we already are from the rest of the natural world. We coevolved with everything around us: "Our ears are attuned by their very structure to the howling of wolves and the honking of geese," writes David Abram. And while it's true that we've already left much of that behind, let ourselves get "caught up in a mass of abstractions, our attention hypnotized by a host of human-made technologies that only reflect us back to ourselves," it's also true that the connection between us and everything else has not yet snapped entirely.[122] Where I live, people are talking about reintroducing the wolf, bringing back that howl. Now, though, we really do seem poised to take the next step in the other direction. Having focused on our own kind to the exclusion of others, we're urged on to a kind of species suicide. Instead of backing down a little, leaving room for the rest of nature, we're firing up the pyre for man as well.

The technoprophets have made a persuasive case—maybe even, in Joy's words, "a realistic and imminent" case—that we will soon be able to leave humanness behind. "There is no need to worry about mere robots taking over from us," writes Brooks. "We will be taking over from ourselves. . . . The distinction between us and robots is going to disappear."[123] In Moravec's words: "I consider these future machines our progeny, 'mind children' built in our image and likeness, ourselves in more potent form. Like biological children of previous generations, they will embody humanity's best chance for a long-term future. It behooves us to give them every advantage and bow out when we can no longer contribute."[124] We would vanish in what the genetic enthusiast Gregory Stock calls a "pseudo-extinction," "spawning our own successors by fast-forwarding our evolution."[125]

If so, these questions of meaning would be left behind for good; there'd be nobody left to worry about work, or challenge, or satisfaction, or sweat. They would simply disappear into the ether; whatever "intelligence" remained would have something else on its mind. Calculating how to speed up its processors, doubtless. But it is an irony, worth noting at least in passing, that the very joy of doing science, which has brought us to this pass, will vanish as completely as the joy of marathon running or potato growing. Science is driven by greed and by ego, of course. (Any doubts on that score should have disappeared when Craig Venter, who was then the CEO of Celera Genomics, revealed that some of the DNA used to sequence the human genome had in fact belonged to him, not the pool of twenty donors from five ethnic groups that was its supposed source.)[126] But it's also driven by the challenge, the satisfaction, the sheer pleasure of the work. Witness Moravec and Kurzweil and Brooks and Joy, happily isolated in childhoods of wonkish abandon. "On January 12, 2002," writes Brooks, "I held a party at my house for all my graduate students. We had champagne and cake, and watched a movie, *2001: A Space Odyssey.*" Brooks had chosen the date because, in the movie version, it's when HAL 9000, the computer star, is first switched on. "That movie, more than any other single event, had changed my life, inspired me as a teenager to dedicate my life to building intelligent machines. I still cannot watch it without my heart quickening and tears coming to my eyes often. It is awesomely inspiring to me."[127]

But scientists, perhaps even more quickly than the rest of us, will find those lines of satisfaction closed when "cyberminds" take over. "As fast as human science is," write Gregory Paul and Earl Cox, "it is still a gradual process of hard-working people of varying competence levels doggedly accumulating data day by day, trying to lead human lives, waiting for that 'aha' of inspiration that comes along once in a great while. The science of intelligent robots will progress thousands or millions of times faster." Where biologists now tromp through the field season after season, "robotic biologists will be able to scan, document, and analyze entire ecosystems in days." Where archaeologists make a life's work out of puzzling over shards and fragments, "figuring out ancient writings will be a snap for cyberminds capable of uploading and deciphering with hyper-sophisticated code-breaking programs in a day."[128] Well, great. But scientists out in the field are among the happiest people I've ever known. Their happiness, and the thousands of other happinesses humans have invented for themselves, would simply evaporate.

Most of the men and women at work on these technologies have no such overarching goals, of course. They are scientists keen for the next small discovery, and businesspeople less interested in enhancing your brain than in enhancing next quarter's bottom line. From such legions comes rapid but incremental change: the circuitry gets a little smaller with each prototype, the design process a little smoother. A guy in a lab in China figures out how to manipulate atoms at only 200 degrees below zero; somewhere in California a graduate student knocks another hundred degrees off; pretty soon it's happening at room temperature. Each of the parts seems like a . . . part. It all happens without some central vision or guidance; the market provides coordination enough.

Still, as we have seen, this future has its prophets, men (mostly) who have looked up from their lab benches long enough to contemplate the future they are helping to usher in. And they in turn are ringed with acolytes, a loose gaggle of science buffs and academics and short-story writers, who have taken their work and turned it into a kind of gospel, adorned with apocalypses and with heavens. They themselves are less important than the Kurzweils or the Moravecs—they're not inventing anything—but the evangelistic note they sound may ring more loudly in the years ahead.

Consider, for instance, Vernor Vinge, a mathematician, computer scientist, and sometime sci-fi author. Sometime between 2005 and 2030, he calculates, we will possess "super-human intelligence," whether by simple programming, or by connecting humans and computers, or by genetic manipulation. And when that happens, "progress will be much more rapid." Humans, he notes, can already "progress" faster than animals because our minds allow us "to internalize the world and conduct 'what-ifs' in our heads," outpacing natural selection. But soon, probably by accident ("we were just tweaking some parameters") our computers will soar past us. "Now, by creating the means to execute these simulations at much higher speeds, we are entering a regime as radically different from our human past as we humans are from the lower animals." That moment, which he calls the Singularity, will mean "a throwing-away of all the human rules, perhaps in the blink of an eye—an exponential runaway beyond any hope of control. Developments that were thought might only happen in 'a million years' (if ever) will likely happen in the next century."[129] Vinge is coy about the timing, but he allows, with a smile, that some of his fans have formed the "2014 club. May thirteenth, 2014." Technologically, these people live at the opposite pole from the fundamentalists with their "In Case of Rapture This Car Will Be Empty" bumper stickers. But not emotionally.

Beyond the Singularity lies—well, the usual laundry list of eschatological joy, of a kind we've already heard. But sometimes the echoes of our cultic past are strong. The futurists are fond of quoting Arthur C. Clarke: "Any sufficiently advanced technology is indistinguishable from magic."[130] They imagine "uploading" themselves into computers, and then, in virtual environments, becoming "anything you like. You can be big or small; you can be lighter than air, and fly; you can walk through walls. You can be a lion or an antelope, a frog or a fly, a tree or a pool, the coat of paint on a ceiling."[131] They imagine, that is, just the sort of trips shamans have been taking for millennia with nothing but the juice of jungle vines.

They even foresee a kind of ethical Singularity, a moment when unhappiness is banished. In his Web-published book *The Hedonistic Imperative,* the "British Nano-Neurology Futurist" David Pearce describes in enormous detail how "nanotechnology and genetic engineering will eliminate aversive experience from the living world. Over the next thousand years or so, the biological substrates of suffering will be eradicated completely" as we strive for "the neuro-chemical precision engineering of happiness for every sentient organism on the planet." In place of the genetically transmitted predisposition to "sadness, anxiety and malaise," it will become "purely an issue of (post)human decision whether unpleasant modes of consciousness are generated in any form or texture whatsoever." "The nature lover will be able to contemplate with awestruck reverence scenes of overpowering sublimity eclipsing the superficial prettiness on offer before," and a musician, with upgraded "functional modules which mediate musical appreciation," will "hear and have the chance to play music more exhilarating and numinously beautiful than his or her ancestors ever dreamed of." Meanwhile the hedonist, relieved by doses of chemicals from the "gene-inspired perversion" of jealousy, will discover with "a whole gamut of friends and lovers" that "what had previously passed for passionate sex had been merely a mildly agreeable piece of foreplay," and "a painter or connoisseur of the visual arts will be able to behold the secular equivalent of the beatific vision in a million different guises, each of indescribable glory."

Ten minutes' thought reveals this as the nonsense it is: if we know anything, it's that art and sex and rock climbing derive much of their glory from the intense concentration that adversity and risk entail, that the ecstatic experience arises more from sensory focus than from sensory overload. The world *falls away.* But this is low-rent religious writing, a rapture as detailed and as tacky as any storefront preacher's. "At some momentous and exactly dateable time," writes Pearce, "the last unpleasant experience ever to occur on this planet will take place. Possibly it will be a minor pain in some obscure marine invertebrate."[132] The lion will lie down with the clam, and they'll do some hits of Ecstasy.

Of course, in many of these visions the inhabitants of earth will have long since lit out for the Final Frontier, anyhow. Artificial intelligences will be prowling the universe, spaceships stored with human downloads going where no man can go. Or perhaps, to quote one professor who has considered the matter, "machine-men may deem it necessary to exterminate men-machines." In this case "human evolution would continue, but only through the further evolution of the triumphant machine-men who would be obliged—if they have any sense of history—to recognize *Homo sapiens* as their progenitor. . . . *We lustful hungry bipeds may simply be nature's way of giving rise to a race of electronic angels* [italics mine]."[133] The physicist Frank Tipler, in his bestseller *The Physics of Immortality,* a book dense with equations and technical appendices, argues that if we do not stand in the way of the superior intelligences now evolving, then we shall "have life after death in an abode that closely resembles the heaven of the great world religions."[134]

This is, in other words, old wine in new skins. Very sweet and understandable human dreams—youth whose bloom doesn't fade, happiness untinged by the suffering of others, love without taint of the impure—that we have contemplated since we began contemplating. What's new is the chance that, in some sense, we will be able to make them come true. Or that they'll seem close enough that we'll try for them, mangling our world and its meaning in the process. And it's interesting, since these prophets have obviously sipped deep from the wells of ecstatic mysticism, how little they seem to have drunk from the streams of common sense that have always coursed through our literature and myth as well. Perhaps childhoods devoted to rewiring computers had no room for such tales, but one wishes that, say, King Midas was as well known as HAL 9000.

For when you speak, as Eric Drexler speaks, of a "genie machine," you're in Midas territory.[135] A great hedonist, King Midas was the first man to plant a rose garden, and when Silenus happened to trample his bushes the king was offered any reward he wanted by the gods. He said, of course, that anything he touched should turn to gold—a good idea, only, like the image of the earth turned to sterile diamond by out-of-line nanobots, too much of a good idea. Sometime after he'd turned his daughter into gold, the gods finally decided he'd learned his lesson and let him rinse off his golden touch in the river Pactolus. (Not that the lesson really took; shortly afterward he screwed up again and the gods gave him the ears of a donkey, an early exercise in transgenic manipulation.) Probably Midas should be our patron saint, reminding us constantly that More and Better are not always found together.

If it's too much to expect the technological prophets to read old myths, they could turn to a source closer to home. The vast literature of science fiction, which once offered a glimpse into a future filled with spaceships and jet cars, now offers a glimpse into . . . hell. Those people committed to imagining the future have taken all the possibilities raised by the new technologies and, thinking them through, have dreamed up a galaxy of dystopias, each more unpleasant than the one before. Forget Jules Verne; it's Aldous Huxley who turns out to have set the style for the genre with *Brave New World,* his 1932 account of a world dehumanized by soma and Centrifugal Bumble-Puppy, a land beyond meaning. Each new technology has spawned its own cautionary literature. Cloning? Read anything from *The Boys from Brazil* to *Lives of the Monster Dogs.* There are subgenres within subgenres; one critic cites four "cautionary novels" from the 1980s alone about human-ape hybrids.[136] Even *Star Trek* couldn't imagine a workable future in a genetically engineered world. One of the recently authorized *Star Trek* novels explains the dominance of mere humans aboard the *Enterprise* by flashing back in time to the Eugenics Wars of the 1990s. "Have the brass at Starfleet lost their paper-pushing little minds?" quips the irrepressible Dr. McCoy. "Human genetic engineering has been banned throughout the Federation since its very founding—and for good reason!" (That reason being Khan Noonien Singh, leader of a gang of supermen reared in an Indian laboratory and then loosed upon the universe.)[137]

Robots? The concept first emerged from medieval Jewish stories of the golem, a giant clay figure that comes to life, invariably to cause trouble for the rabbis who built it. The word "robot" was coined by Karel Capek in his 1920 play *R.U.R.*, which ends only "when all humans have been eradicated."[138] Isaac Asimov did his best to control the creatures by formulating his Three Laws of Robotics, which tried to erect a logical wall high enough to keep robots from hurting humans. It was a constant struggle. But less hopeless, perhaps, than the lives depicted by William Gibson in his various cyberpunk novels, which anticipated the coming of everything from neural implants to black-market genetic materials. "Gritty" is how they're usually described, but that's like calling an ice age chilly: Gibson is the Dante of the coming age, describing the circles of the Inferno in advance of their creation. There are also nanotech dystopias (Neal Stephenson's *The Diamond Age,* Michael Crichton's *Prey*) and AI horrors: it's hard to imagine that anyone who watched Keanu Reeves struggle to escape the Matrix would long for the development of superhuman intelligence. From *Johnny Mnemonic* (Keanu Reeves again, this time involved in a "gritty union of biology and technology . . . filled with jarring, violent landscapes of wrecked cities, cyborgs, and hypertechnology run amuck") to the latest novels about "brain uploads" (Greg Egan's *Permutation City,* a story of epic desolation), there's hardly a happy ending imaginable.[139]

The *New York Times* writer Patricia Leigh Brown noted in a 2002 essay that science fiction writers have it increasingly tough because the labs spin out marvels and monstrosities at such a pace that the novelists can't write fast enough to keep up. Bruce Sterling once wrote about goats genetically engineered to produce plastic explosives; earlier this spring Canadian researchers managed to implant spider genes in goats, enabling them to produce milk containing superstrong silk. The "tooth phone" was unveiled earlier in the year—top that, Q. "In Australia, researchers in quantum optics say they have 'teleported' a radio-signal message in a laser beam, using the same kind of principles that enabled Scotty to beam up Captain Kirk."

The real problem, however, is less the pace than the power of these new inventions. If, as a writer must, one sits quietly and tries to imagine the actual use of such devices, the imbalance between technology and humanity becomes too clear. Making *anything,* knowing *everything*—these sound like dreams. But if you feed them through a creative mind they quickly turn out to be nightmares. Sterling reported he had recently returned from a conference "where computer scientists hobnobbed about genetic algorithms and ubiquitous computing." He came away, he said, with the germ of a novel, about "a world in which every object is seeded with sensors, where black helicopters hover over smoking ruins and spew out computers that detect breathing."[140] More becomes too much; it overmasters and annihilates, removing possibility instead of creating it.

Though it lacks the latest in technology, and though its settings are nearly grit-free, Arthur C. Clarke's *The City and the Stars* takes these questions as deep as anyone's taken them. Written in the 1950s, it tells the story of a young man named Alvin, "born" into the city of Diaspar. But "born" is the wrong word, for Diaspar is a city of immortals; they are stored in the city's vast computer banks and then, every few thousand years, reanimated to live long lives before they are stored away again. Long lives, and rich lives:

> There were a million things to occupy their lives between the hour when they came, almost full-grown, from the Hall of Creation and the hour when, their bodies scarcely older, they returned to the Memory Banks of the city. In a world where all men and women possess an intelligence that would once have been the mark of genius, there can be no danger of boredom. The delights of conversation and argument, the intricate formalities of social intercourse—those alone were enough to occupy a goodly portion of a lifetime.

And, of course, they have hobbies—art, dance, perhaps woodworking with tools that massage your hands as you saw and plane.

And yet Alvin, alone of his race, feels compelled to escape—to tunnel out of the domed city, and find on the other side a small colony of people dwelling in the open air. They, too, are wise, but they have chosen mortality, reality, life. (In my favorite scene, Alvin takes his first hike, feels his lungs and legs really working for the first time; that night, again for the first time, he sleeps.) Alvin wanders around their village, his mind hardly able to take it in. He sees an old man (though barely a hundred!), "his hair completely white and his face an unbelievably intricate mass of wrinkles": "He seemed to spend most of his time sitting in the sun, or walking slowly round the village exchanging greetings with everyone he met. As far as Alvin could tell he was completely contented, asking no more of life." And Alvin sees children, "those little creatures who were as strange to him as any of the animals. . . . And yet while they baffled him, they aroused within his heart a feeling he had never known before. When—which was not often, but sometimes happened—they burst into tears of frustration or despair, their tiny disappointments seemed to him more tragic than Man's long retreat after the loss of his galactic empire. . . . He was learning tenderness."[141]

You could call that sentimental, and you could also call it true. If the human scale is about to be overwhelmed, we've at least been warned. These new technologies, as I said earlier in this chapter, are visions. You'd be tempted to say they are "science fiction" visions, but that's wrong. They are *science* visions, and in some form or another they will come true. Maybe not quite as fast as the plans of the genetic engineers, but fast enough. The only real hope of heading them off is with some other vision. With some other account of who we are, and what we might be.

CHAPTER THREE

Enough?

We need to do an unlikely thing: we need to survey the world we now inhabit and proclaim it good. Good enough. Not in every detail; there are a thousand improvements, technological and cultural, that we can and should still make. But good enough in its outlines, in its essentials. We need to decide that we live, most of us in the West, long enough. We need to declare that, in the West, where few of us work ourselves to the bone, we have ease enough. In societies where most of us need storage lockers more than we need nanotech miracle boxes, we need to declare that we have enough stuff. Enough intelligence. Enough capability. Enough.

That's the hinge on which this argument turns. I have no shiny new vision to compete with the futurists who dream of making us "posthuman." We need, instead, a new way of looking at the present. If we can come to see it as sufficient for our needs, then perhaps we can figure out how to avoid these new technologies and the risks—physical and existential—that they pose. Perhaps we can find, instead, some conserving instinct within us that lets us stand pat. We'll get to the specifics of how we might do such a thing later, for clearly it will be difficult to blunt our technological momentum. But first we need to answer the novel question of whether we really do pass muster in our present form.

The technological visionaries shout *No!* to that question; in their eyes, we are deeply flawed, beginning with our very bodies. Consider Robert Ettinger, the world's most influential cryogenicist, who said early and often that genetic engineering and other new technologies would usher in a "golden age." One prominent feature of which would be, in his opinion, the "elimination of elimination." If, he reasoned, "cleanliness is next to godliness, then a superman must be cleaner than a man. In the future, our plumbing (of the thawed as well as the newborn) will be more hygienic and seemly. Those who choose to will consume only zero-residue foods, with excess water all evaporating via the pores. Alternatively, modified organs may occasionally expel small, dry compact residues."[1] Ettinger had trouble with other orifices as well: one of his friends had pointed out to him that a "multi-purpose mouth" was "awkward and primitive" to the point of "absurdity. An alien would find it most remarkable that we had an organ combining the requirements of breathing, ingesting, tasting, chewing, biting, and on occasion fighting, helping to thread needles, yelling, whistling, lecturing, and grimacing."[2]

Ettinger was not alone in his self-loathing. It permeates this subculture, the constant lament that at best people resemble, say, Yugos—decent basic technology at a reasonable price, but nothing to get fixated on. To a technician's eye, their defects are simply too annoying. Hans Moravec once reflected on an Asimov short story about an android who wanted to become a real person. "That's a cute story," he said. "But I read it and I thought, *Why in hell do you want to become a man when you're something better to begin with?* It's like a human being wanting to become an *ape.* 'Gee, I really wish I had more hair, that I stooped more, smelled worse, lived a shorter life span.'"[3]

Some people, admits the AI pioneer Marvin Minsky, seem to "like themselves just as they are. Perhaps they are not selfish enough, or imaginative, or ambitious. Myself, I don't much like how people are now. We're too shallow, slow, and ignorant."[4] As a species, he notes, we "seem to have reached a plateau in out intellectual development. There's no sign that we're getting smarter . . . has any playwright in recent years topped Shakespeare or Euripides? We have learned a lot in two thousand years, yet much ancient wisdom still seems sound, which makes me suspect that we haven't been making much progress." Basically, he maintains, it's a hardware problem: citing a Bell Labs study, he calculates that humans can learn and remember only about two bits per second. So even if you did nothing but learn twelve hours a day for a hundred years, the total would only be about three billion bits—"less than what we can store today on a regular five-inch compact disk."[5]

"There's got to be a better way," implore Gregory Paul and Earl Cox in their book *Beyond Humanity.* "Why do we have only one heart, why not two in case one fails? Or two livers?" We evolved to be "gatherer-hunters doing a better job of picking up nuts and berries"; as it turns out, we were also "marginally adapted" for high-level physics and novel writing, just as archaeopteryx, the first dinosaur with wings, managed "barely adequate flight." But in essence we're a failure: even "the erect bipedal posture of which we are so proud makes us so unstable that falling on flat ground can have devastating results."[6]

Freeman Dyson, the eminent physicist and futurist, told of the time his five-year-old first saw him naked. "Did God really make you like that?" she asked. "Couldn't he have made you better?" The "only honest answer," he said, is, "of course, yes."[7]

But it's not just our bodies and minds that seem impossibly crude. When they look at the societies those bodies and minds have built, the visionaries despair as well—they sound, in fact, like over-the-top Hobbesians. George Gilder, usually described as a "high priest" of Silicon Valley for his influential articles and seminars, says only wave after wave of high-tech change can save "millions or even billions of people from their immemorial fate as members of a barbarian mob, plunged in ignorance."[8] On the bulletin boards of the Web, you can find zealous "transhumanists" not just explaining how the world will evolve past our species, but also debating questions like whether it's morally necessary to kill "Luddites" who stand in the way of such progress: "If we don't establish a transhuman future, then Planet Earth is doomed to a rather dismal malthusian extermination from disease, hunger, and lack of resources. If we don't build a transhuman future, billions of people WILL die. . . . What do you intend to do about it? To what ends are you willing to save billions of lives? What means are you willing to accept to achieve those ends?"[9]

If we start to think this darkly, then it's only a matter of time before the tech prophets carry the day, if only because existential despair makes it hard to shake off the gloom and put up a fight. In 1998, French writer Michel Houellebecq published *The Elementary Particles*, a runaway bestseller in his home country; it chronicles the rise of a posthuman species that has transcended mankind's tragic flaws. "There remain some humans of the old species," he writes. "At present their extinction seems inevitable. Contrary to the doomsayers, this extinction is taking place peaceably, despite occasional acts of violence, which also continue to decline. It has been surprising to note the meekness, resignation, perhaps even secret relief with which humans have consented to their own passing."[10]

But how is it that we've reached this point? Environmentalists share, I fear, some measure of the guilt. The movement to value everything else on earth has often talked carelessly about people, spreading the idea that we are a grim and uncontrollable race, a cancer cell metastasizing unchecked across the defenseless fabric of nature. From the moment that the Reverend Malthus first advanced his theory that reproduction would inevitably outstrip food production, a certain kind of despair has informed an awful lot of what we would eventually call environmentalism. Whenever I've given a lecture on some of the ways we might mend our environmental troubles, someone from the audience has usually risen to ask if, say, global warming isn't simply a way for nature to "get rid of us," a species more trouble than we're worth. I feel that despair myself sometimes; there are days when my own consumer lust and essential apathy convince me we're doomed.

That mood leaves one vulnerable to the siren song of posthumanism—to the idea that we should be radically reconfigured or, better yet, surpassed by some wiser race. To the idea that new technology will save us from ourselves. "Nanotechnology will reverse the harm done by the industrial revolution," promises the Nobel laureate Dr. Richard Smalley, who leads a nanotech research team at Rice University.[11] "Commonplace notions paint a future full of terrible dilemmas," writes Eric Drexler. Instead, molecular engineering will give us "green wealth unfolding from processes as clean as a growing tree."[12] Environmentalists have written movingly about the dangers of industrial agriculture, about the habitat destroyed by our fields, about the species driven extinct by our expansion. So how to argue with those who want to "save" species by cloning them, or who imagine nanotech manipulation replacing farming?[13] "Humanity will become a low-pollution system largely decoupled from terrestrial nature," exults one writer. "Bison will roam the wide open plains again."[14] If every prospect pleases and only man is vile, then perhaps the answer really is to take man out of the picture. Usher in the bots on a cloud of green.

But there's another possibility. We could drop the misanthropy and look at ourselves with clearer eyes. Yes, we've damaged the environment, we've enslaved our fellow man, we've slaughtered on a vast scale. This is the truth. But it's not the whole truth, or even the main one. To put it bluntly, *the fight to ward off a posthuman future begins with at least a muted celebration of the human present.*

People are okay. I'm okay, you're okay, the lady who stands there forever talking with the cashier at Shop'n Save is nonetheless, deep down, okay. Maybe even more than okay—as birds have been blessed with flight, we have been blessed with an exuberant consciousness that has given rise to much good. So much more good than bad. The God of Genesis looks around the newborn creation and pronounces it *all* good. Not just the sun and the stars, not just the plants yielding seed and the trees bearing fruit, not just the "great sea monsters" and "everything that creeps upon the ground." Us, too. Even though we've done enormous damage. Even though we use the same mouth to eat and kiss.

And there's more. We're capable of getting better still, all on our own. We're capable of the further transformations necessary to redeem the world. Humanity is not, in the words of one techno-prophet, "running out of steam . . . burdened by deep-set flaws that will always limit our minds' potential."[15] We've made a hash of the world in many ways, but the damage is not beyond repair. We're close enough to the good life that we don't require a magical genetic rescue, a deus ex nanomachina, a fairy-tale ending full of moral robots.

Slowly, but with reasonable steadiness, we've made the world less . . . Hobbesian. Legalized slavery has been scrubbed from most of its corners; we now indict as international war criminals those who rule by the normal methods of an earlier day. Most of the Western world has dispensed with capital punishment, and even in the United States we seem to weary a bit of its use. In the lifetimes of people not yet old we have seen the end to official racism, and the galloping emancipation of women in cultures across the globe. The "handicapped" are brought more fully into the life of our society; difference of every kind, in fact, is more accepted and celebrated. It's been a bloody century we've endured; in the wake of World War II, the philosopher Robert Nozick made what he called "the argument from Hitler"—that the Holocaust demonstrated it would be no "special tragedy" were man to disappear.[16] But you could as easily make the argument from Eisenhower: that millions were willing to fight against that tyranny, just as people fought to upend the bleak Communist totalitarianisms. I remember going to a genetic engineering conference in Boston the week after September 11, 2001, when it was awfully hard to think about anything but those airplanes. "We are a pretty awful species," said one of the speakers. "There are times when I'd like to see us extinguish ourselves." Who, in those bad days, could argue? And yet, in the face of that evil, people by the hundreds of thousands were at work trying to heal the damage. In the weeks and months that followed, the *Times* printed one obituary after another—of dishwashers, of bond traders—and they were stories of people with rich, full, caring, useful lives. The obituaries were a revelatory look at what our neighbors were actually like.

Even with our environmental problems, perhaps the most daunting of our woes, we've begun to see the solutions. In the West, our air is a little cleaner, and we can swim in our rivers again. Wind is now the world's fastest-growing source of electrical power, expanding fourfold from 1995 to 2000; scientists have demonstrated that soon it will be possible to use that energy to produce hydrogen fuel for our vehicle fleets.[17] If we set our minds to it, such technologies could grow exponentially, as computer power has: the environmental economist Lester Brown calculated in 2001 that if we double the world's wind-generating capacity each year for the next decade, by 2010 we'd be producing twice as much electricity as the world consumes, just from spinning windmills. The change won't happen that fast without an unprecedented investment of money and will, but an energy revolution is clearly "under way. The only questions are how rapidly it will unfold, whether it will move fast enough to prevent climate change from getting out of hand."[18] If governments sent signals to the markets so that they calculated ecological damage into prices, the change could come as fast as any described in this book. And technologies like wind power come with no unmanageable downside, no great risk to anyone but the owners of coal mines and oil wells. Sensibly placed, they would, if anything, add meaning, not subtract it. A spinning windmill is a lovely sight, breeze made visible.

Dealing with our environmental problems means, above all, changing the trajectory of our various curves—for example, the curve of carbon dioxide emissions, which has been shooting up at the same exponential rate as computer-chip capacity. If you want proof that it can be done—if you want the single best statistical answer to despair, the single best proof of the human capacity to limit itself—then consider the curve of human population.

For a very long time now, we've thought of fertility as the root problem. We've feared the population bomb, imagined a world grown so crowded we were each standing on our own few feet of soil. And population was climbing ever upward, a classic spike. But then some people took it upon themselves to try to slow that curve down. Beginning in the rich countries, environmentalists and development experts started trying to spread birth control around the world. Soon, local people in developing countries picked up and improved on their efforts; educating women, they realized, was the key to reducing family size. And it is working. The world's population is still growing, because of the incredible numbers of people now entering childbearing age. But the typical woman is now having far fewer children, from an average of more than six a generation ago to an average of fewer than three today. Demographers expect that number to keep dropping, to below the replacement level of two kids per mother. As a result, the world's population won't double even once more; as early as 2070, by the latest predictions, the number of human beings may actually begin to fall.[19] The tunnel is still curving upward, but the curve is starting to flatten, and with the periscope of statistics you can make out the light at its end. In the meantime, of course, the earth will strain to support those new people. When the peak finally hits we'll face new problems of a graying society. And there's still the steeply climbing consumption curve to worry about. But human beings—with education, with effort, with wisdom, with bittersweet maturity—have restrained their own population growth. They've taken away the trump card of every technologist who insisted we had no choice but to press forward forever against the demographic tide. In this one area, we've said, "Enough." Late in the game, more slowly than we should have—but still, *enough*.

Is it possible that we might want to say the same thing about these dawning hypertechnologies?

Might we, that is, conclude that we have sufficient power now and do not need or want another quantum increase? That while the vast majority of scientists and engineers should keep at their benches working small wonders for us, we might ask them to steer clear of new technologies so mighty as to change the essential nature of our bodies or our lives? Is it possible we're at an enough point?

It's a challenge even to think such heretical thoughts, for we're used to bowing and scraping before the engineers. We consider all the blessings that technology has brought us and imagine it would be ungrateful to question it. There's even a standard rhetoric, used by all the futurists because it resonates so strongly with us:

Lee Silver: "Two hundred years ago, people said, 'Well, some people just get infected and die and other people don't, and that is just the way it is. That is the way God intended it to be."[20]

Ray Kurzweil: "How many people living in 2001 would want to go back to the short, brutish, disease-filled, poverty-stricken, disaster-prone lives that 99 percent of the human race struggled through?"[21]

Hans Moravec: "During the last three centuries in the industrialized countries, slave and child labor and hundred-hour factory work weeks have given way to under-forty-hour weeks and mandatory retirement."[22]

Such words come with an implied threat: if we don't keep going forward, then we will retreat into the pit. It's either human enhancement or slave labor. I got a letter recently from someone outraged by an op-ed piece I'd written: "Those who proclaim they don't want to tinker with nature should forgo all antibiotics and aspirin." But this is just rhetorical sleight of hand. No one I've ever met proposes that we stop treating infections (in fact, the people who do treat infections want us all to stop using antibiotics for trivial purposes, because the number of drug-resistant bacteria is on the rise). Absent advanced robotics, we will not return to hundred-hour work weeks. Keep your Tylenol. *We have reached a point of great comfort and ease relative to the past; the real question is whether, having reached that point, we want to trade it in for something essentially unknown.*

And in making that decision, the history of what technology has done for us in the last two hundred years is largely irrelevant. As the mutual fund ads always say in small print at the bottom, "Past performance is no guarantee of future results." What liberated in the past may imprison in the future. In this case, as we have seen, the engineers probably can deliver: they can speed things up, they can "enhance" our minds and bodies, they can bury the patent office in new inventions. Such "improvement" may or may not be physically dangerous; it may or may not unleash designer germs and gray goo. But I've tried to show why I think it will make life meaningless.

We are leaping across thresholds. While the jump to microscale technology may have made life easier, the further jump to nanoscale engineering will eventually drown us in a gushing cornucopia. While the jump to modern medicine may have freed us from many ills, the next leap to human genetic manipulation will imprison us in a house of distorting mirrors.

That's how thresholds work: up to a certain point something is good, and past that point there's trouble. One beer is good; two beers may be better; eight beers you're almost certainly going to regret. If you drive your car at 55 miles per hour, you'll get where you're going faster than if you drive it at 20 miles per hour—but if you drive it at 155 miles per hour, odds are you're going to die in a ditch. A certain amount of carbon dioxide in the atmosphere helps keep our planet habitably warm, but now we're spewing so much of it from our tailpipes and power plants that we threaten the earth's equilibrium.

Less isn't always better; there's no need to reject "the Enlightenment" or "Western civilization." But more isn't always better, either, though that's the simpleminded assumption that drives these clever new technologies. Having a good memory is better than having a poor one, so having a perfect memory would be best of all—let's stick a chip in there, or a new gene. But perhaps being able to forget is one of the great gifts we've been given, by evolution or by God or by great good luck.

Judging when you've reached this "enough point" is, admittedly, no easy trick. You might stop short and miss some real improvement; you might overshoot and hit some wall. Is the fifth beer the one that's going to make you feel *really* good, or the one that's going to make you puke? (And as with beer, so with technology—once you're drunk or bedazzled, it's harder to resist the next step.) It's always going to be a guess, a question of feel, an art: Have I held this note long enough? Is this brush stroke wide enough? But there are plenty of clues to alert us that we're near a technological saturation point, past which we will hit radically diminishing returns. The sheer speed of the world, to name the most obvious example, overwhelms our ability to keep pace; we feel a kind of frantic restlessness, which we sense will be alleviated only by slowing down. Such notions began to emerge in the 1960s—"Small is beautiful"; "Drop out"—but progress still seemed to be producing enough of a return, and by and large we reenlisted for another tour of duty. (In fact, as hippies became boomers they turned ever more enthusiastic.) But now the same questions are in the air again, because the world seems about to run away from us. Our food has been genetically modified, which makes us uneasy; our children are about to be, which should make us cringe.

With some technologies, we can already draw the line: Germline genetic engineering would be a mistake. Nanotechnology and robotics may not have matured far enough for it to be obvious exactly when they'll cross a similar threshold. But anything with the power to take us "posthuman" should be watched with a beady eye; each incremental advance should be presumed dangerous until proven otherwise.

In the shadow of such fears, we may finally start to find the idea of "enough" more palatable than we have in the past. Right now our technology is advanced enough to make us comfortable, but not so advanced that it has *become* us. We have enough insight from Darwin and Freud and Watson and Crick to allow us to understand some of what drives us, but we're not yet completely reduced to hardware. We have enough Prozac for the incapacitated and pain-ridden, but it's not encoded in our genes. We have enough medicine to give most of us a good shot at a long life, but not so much as to turn us into robots. We are suspended somewhere between the prehistoric and the Promethean. Closer to the Promethean. Close enough.

Here's one way to try to make this judgment call: Can you think of aspects of your daily material life that would be dramatically improved by the next dose of technology? Changes that would matter the way electric light mattered, or indoor plumbing? That would be worth the various risks I've described? When the technologists try to sell the future, it's those day-to-day improvements they usually fix on. Rodney Brooks, interviewed in the documentary *Fast, Cheap & out of Control,* was asked for an example of how the world will be better when we have "embedded robots." He answered as follows: "I get annoyed at all the fluff that accumulates on my TV screen. The dust. Well, what if you could buy twenty robots for a dollar in a little bottle, empty this bottle onto your TV screen, and while it's on they absorb electrons from the screen. When you turn it off, they each scurry out and scrub a corner of the screen and then scurry off to the corner again to sit there. It sounds crazy by today's standards, but that will be a cost-effective way to clean your TV screen."[23] No more the semi-annual four-second wipe with a damp rag; instead, a bottle of robots.

"Inside the home," explains H. Keith Henson in a volume published by MIT, "there are countless applications for trivial nanotechnology." A self-cleaning rug "would ripple like cilia in your lungs to move all the little stuff that fell on it to a central bucket. . . . A still later version might deal with the classic clothes-on-the-floor problem that has broken up so many marriages. Left for more than a few minutes, underwear, socks, even dress clothes could be disassembled by the carpet and rebuilt in dresser drawers or hanging in the closet."[24] Twenty years from now, says Brooks, when we have neural connections from our brain to the Internet, "we will be able to think the lights off downstairs instead of having to stumble down in the dark to switch them off," and in the morning "we will be able to think the coffee machine on, down in the kitchen."[25]

Eventually, says Mary Walker, of IBM's home automation division, "smart ID" chips will be implanted inside you. Then "your body temperature might give your stereo system cues as to your mood and it would select appropriate music"; the chip could also "compute how much of your body weight is fat, and offer suggestions for diet recipes to the refrigerator."

"The whole table-setting and -clearing problem will require a completely new way of thinking," explains Brooks.

> Perhaps in the future our dining room tables will also be the place that we store all our dishes when we are not using them, in a large container under the tabletop surface. When we want to set the table, small robotic arms, not unlike the ones in a jukebox, will bring the required dishes and cutlery out onto the place settings. As each course is finished, the table and its little robot arms could grab the plates and devour them into the large internal volume underneath. With direct water hookups into our dining rooms, much as we have them now in our kitchens, the dining room table could also be the dishwasher. It would wash the dishes after each meal and leave them down in its recesses, waiting for the next meal at which they are needed.

Also, "robotic magnifying mirrors" will soon be able to "follow us around the bathroom, so that whenever we are ready to use them, they will already be positioned at the perfect angle."[26]

At a certain point, says the philosopher Albert Borgmann, technology "mimics the great breakthroughs of the past, assuring us that it's an imposition to have to open a garage door, walk behind a lawn mower, or wait twenty minutes for a frozen dinner to be ready."[27] Borgmann wrote that twenty years ago. Now the salesmen insist that it's an imposition to have to push the button that opens the garage door when the smart ID chip embedded in your bicep can do it for you. That seems to me a reasonable definition of the enough point. And perhaps not just for me. Since the mid-1950s, pollsters have annually asked Americans if they are happy with their lives. The numbers who say yes have declined slowly but steadily for four decades, even as technology has dropped more and more conveniences from the sky. The researchers have found that people expect material progress to increase, and also expect "inner happiness" or "peace of mind" to decrease. "The results of such surveys indicate that in fact a substantial majority of people believe there is a negative correlation between progress and happiness."[28]

At some spot in the past, that wasn't the case; if you'd been carrying water in a bucket from the stream, then an electric pump must have seemed like one swell deal. But we're no longer at that spot in history. We have running water, and it takes very little effort to reach out and grab the faucet handle with our fingers and twist it on. I like refrigerators, but I don't need mine calculating my body fat and suggesting diets.

With average daily life in the Western world, then, I think we're quite near the enough point. But "average daily life in the Western world" leaves quite a few holes. It's entirely possible that, despite our general satiation, there are emergencies around us so dire that they demand that we develop these new techniques.

As I've talked to technologists, and read the massive literature, three such emergencies have risen to the top, realms where it seems like we might legitimately need a quantum leap in technological power. Two are real problems: helping medicine deal with illness, and aiding the vast numbers of poor around the world. The third is a deep human dream: conquering death. We shall examine them in turn, trying to figure out whether they constitute sufficient reason for forging ahead, because these three questions are central to figuring out whether indeed the world is good enough now. In the words of the genetic engineering proponent Gregory Stock, "A few romantics may one day look back on our era as a golden one, but far more future humans will see today as a primitive, difficult time, far inferior to the world they know."[29] Perhaps so. Perhaps we are really leading primitive and difficult lives. If that turns out to be true, then the enough point waits far in the future.

• • •

Purely genetic diseases number about seven thousand; a few, such as cystic fibrosis and sickle-cell anemia, are fairly common, but most are extremely rare.[30] Such diseases appear in about 2 percent of all live births.[31] Many more of us—all of us, actually—carry a genetic predisposition to at least one *something* that might conceivably cause us harm at some point. Taken together, these immediate and vague threats to our health are sometimes seen as the principal justification for beginning to manipulate human genes. Adding IQ points will come a little later, the proponents say; at the beginning we want to treat illness. This is, if nothing else, a good rhetorical strategy—laypeople usually defer to doctors. In the words of James Watson, arguing for germline engineering, "I think we can talk principles forever, but what the public wants is not to be sick. And if we make them not be sick they'll be on our side."[32] So we need to know: does the treatment of illness require us to open ourselves to the enormous upheavals that will come with genetic manipulation?

Recall first the distinction between somatic and germline genetic treatment. Germline manipulation is the frightening variety—the making of changes in human embryos that could alter everything from health to IQ, and that would be passed along to future generations. This is a technology that by my lights lies on the far side of the enough point.

Somatic gene therapy, on the other hand, causes less apprehension; it's an extension, albeit a potentially very powerful one, of the kinds of medicine we're used to. Doctors inject healthy human genes into a living patient, in the hope that they can get those genes to take over for the ones producing diseases. If your lung cells, for example, contain a gene that produces an aberrant protein, doctors could attach copies of the "normal" gene to harmless viruses and let the viruses penetrate the cell walls and nuclear membranes of your lung cells. The new genes would be released into the cell nuclei, and with any luck they'd start producing the normal protein.[33] "The body is missing something, and the doctors try to add it," writes Barbara Katz Rothman. "It builds on the traditional Western notion of the body at war against disease, with the doctor adding reinforcements on the side of good against evil, whether the new troops consist of aspirin, cough medicine, L-dopa, or a stretch of DNA. The story stays the same."[34] Or, as Stuart Newman said as we sat in his lab at New York Medical College, "If you have a person who's with us here and now, and he's sick, we accept all sorts of heroic gestures. It's hard to argue you shouldn't give an injection of genes."[35]

For a while, it looked as if somatic therapy was a good idea that wouldn't pan out. More than a hundred experiments produced no clear successes; the new genes didn't make it into the cells in sufficient numbers, or they didn't produce the needed proteins in sufficient quantities. At least one research volunteer, eighteen-year-old Jesse Gelsinger, died when the theoretically harmless virus that the new genes were riding on touched off a fatal immune response.[36] But then, in the spring of 2000, French scientists announced that they had successfully treated three infants who had been born with severe combined immune deficiency, or "bubble-boy syndrome." The disease leaves victims without a complete immune system, easy prey to any germ; untreated infants need to be kept within a sterile bubble or they will die before their first birthday. But Dr. Alain Fischer, a professor of pediatric immunology in Paris, took blood cells from the bone marrow of his infant patients, mixed the cells with a virus containing the replacement gene, and returned them to the bone marrow, where the virus began to proliferate, building immune systems for the children. "This would probably be the first example in any disease where gene therapy could be called a fully successful treatment," said Dr. Michael Blaese, who had worked on the first somatic therapy trials a decade earlier. "You can't distinguish these patients from normal."[37] A few weeks later, another team reported that hemophilia patients in Philadelphia had begun to produce clotting factors in their blood after receiving a virus that contained genes to produce the necessary protein. "Hemophilia may be within a short time of being cured by this approach," said one immunologist. As researchers find new viruses to carry genes into cells, and develop techniques that allow for the direct injection of new genes without any viral vectors, most think the track record will continue to improve. "It's finally coming together," said Dr. Savio L. C. Woo, president of the American Society of Gene Therapy.[38] Not that it will be easy; in the fall of 2002, doctors reported that two of the "cured" immunodeficient infants had developed cancer as a result of the treatment. The new genes "had inserted themselves, backward, inside a gene known to cause leukemia." Federal officials shut down twenty-seven of two hundred gene therapy trials currently under way.[39]

Most news reports about such successes don't distinguish between somatic and germline manipulation—both are "gene therapy," at least in the headlines. And there are researchers who want to use the momentum from the somatic successes to cross the line and begin the germline work that is, for the moment, prohibited. W. French Anderson, for instance, who did the earliest somatic therapy trials, has applied to the National Institutes of Health for permission to begin somatic trials on human fetuses, a procedure he acknowledges would have a "relatively high" potential for "inadvertent transfer to the germline."[40] Many see Anderson's proposal as an attempt to knock down the walls against germline therapy, "a deliberate attempt to breach the germline barrier."[41] So far the government has not said yes—but, unlike many European governments, it hasn't closed the door on the possibility of germline intervention either. And the drumbeat, among those who want to cross this line, continues: germline therapy would be more "efficient" than somatic treatment, more powerful.[42]

Why not simply let medical researchers use germline therapy to treat disease? Why not allow them to expand, from real patients with real problems to embryos that might have problems someday? Tell them to stay away from IQ, leave memory alone, don't start producing designer fullbacks—but if someone is at risk of getting cystic fibrosis, then go ahead and fix the embryo.

The answer is that once you cross this line, there's no stopping.

This is no slippery slope—this is the Khumbu Icefall, the Olympic bobsled track, the double black diamond run from the top of the highest chairlift. "If you develop the technology for putting in genes that enhance health characteristics, then that very same technology can be put in to give a child other kinds of non-health characteristics, like increased talents . . . or increased memory abilities or cognitive skills," says Lee Silver.[43] The line between "cure" and "enhancement" is so blurry as to be no line at all. "If you say you're going to use it for disabilities, but you are not going to use it for character traits, then is schizophrenia a disability but depression a character trait?" asks the bioethicist Adrienne Asch. "When is dwarfism a disability, and when is being short a social problem?"[44] Once it's accepted that parents have a right to use germline intervention to change a predisposition to shortness into a predisposition to average height, asks Richard Hayes, a critic of germline manipulation, how "could you argue that they didn't have a right to predispose their child towards above-average height?"[45] Is the federal government going to start setting height limits?

It's not just the critics of germline engineering who see this slick slope looming; those who see it as inevitable also frankly admit that today cystic fibrosis, tomorrow IQ. "It is impossible to draw the line in an objective manner," writes Lee Silver. "In every instance, genetic engineering will be used to add something to a child's genome that didn't exist in the genomes of either of its parents. Thus, in every case, genetic engineering will be genetic enhancement."[46] Silver's book ends with a chapter of science fiction; his alter ego, Albert Varship, looking back from 350 years into the future, laments the division of *Homo sapiens* into the GenRich and the GenPoor and tries to imagine how it might have come out differently. "The original practitioners drew a moral line between preventing disease and enhancing characteristics. How could anyone argue against preventing childhood disease? But it soon became clear that the moral line was an imaginary one. It was all genetic enhancement. It was all done to provide a child with an advantage of one kind or another that she would not have had otherwise."[47]

In other words, germline engineering for any reason lies on the other side of the enough point. If we adopt appropriate safeguards, we can do somatic genetic therapy, even though it's clearly novel and "high-tech," without leaving behind the world as we've known it. We still live on Earth 1, we're just a little healthier. But if we start manipulating the germ cells of embryos, then we escape the gravitational pull of history and evolution and speed off into the dangerous and demoralizing wastelands of Earth 2, the world beyond meaning.

If somatic and germline manipulation lie on opposite sides of the line, then "therapeutic cloning" lies squarely atop it, a technology that could go either way, and one that we should discuss briefly, because it's been the center of much political debate recently and helps set the tone for this entire discussion. Briefly, researchers want to get their hands on "embryonic stem cells." When an embryo is just a few days old, its cells have not yet "decided" whether to be blood or brain or skin; they're just cells. If you have some, they can be grown in a lab and perhaps be converted into an "unlimited supply of tissue for transplant: new heart muscle for heart-attack survivors; insulin-secreting cells for diabetics; neurons to treat those suffering from spinal-cord injuries, the effects of stroke, or Parkinson's disease."[48] No one knows how well such treatments will work, but no one has any problem with them, either; the use of stem cells is an obvious extension of medicine as we've known it.

The problem comes from how you *get* the cells, which, remember, are contained in a five-day-old embryo. So far, researchers have used leftover embryos from in vitro fertilization clinics. There isn't an endless supply of those, however. And it would be even better to have stem cells that came from the patient, so that when he or she grew the new tissue there'd be less chance of rejection. But your patient is grown up, so how do you get cells from his embryo? The answer is, you clone him. You take a few cells from, perhaps, his skin, extract their nuclei, introduce them into a human egg, and sit back to watch it develop. But only for a little while. As soon as it's had four or five days to divide and grow—as soon as, say, it's 150 cells in size—you go harvest those fetal stem cells, and use them to regrow, say, heart tissue.[49]

Abortion opponents and others who are convinced that an embryo is a person object to that procedure because they see it as creating a life and then, in essence, killing it to help someone else. "Embryo-farming," President Bush called it; a slap at human dignity.[50] Nonsense, say the scientists: we stop the development of the embryo so early that it's only a "blastocyst," not a person. Take your pick; this fight's been under way since long before *Roe v. Wade.*

But when the issue came before Congress last year, anti-abortion activists weren't the only ones who wanted a moratorium on embryonic cloning. So did a pretty good sprinkling of feminists, environmentalists, and other liberal activists—and for an entirely different reason. They worried that once you had taken Fred's skin cells and used them to create a clone so you could grow Fred's new heart muscle, there was no guarantee that that's where you'd stop. You could, instead, implant the cloned embryo in a woman's womb and, voilà, you'd have Fred 2, the first clone. As one witness testifying on behalf of the Justice Department noted, fertility clinics "routinely transfer embryos from the laboratory to women's wombs." Short of posting an FBI agent with a microscope in every IVF clinic in the country, he said, "it will be impossible to stop scientists from using the cloned embryos to impregnate women."[51]

Human clones lie on the far side of what I've been calling the enough point. They are children born with an even clearer road map of their future than the products of genetic manipulation, because in some sense their life has already been lived. Obviously they'd be different from their genetic doubles: they'd grow up in a different time; their environment would push them in different directions. But they would never have the sense of being their own person; they would be designed in a deeper way than any human has ever been designed to date. The first clone will be the first "person" with just one parent, and that parent will in some sense be himself. Virtually every objection to the genetically enhanced child that we discussed in chapter 1 applies to the clone as well.

Not only that, but the first clones would break down the door to all the other forms of genetic enhancement. The technology of cloning is much the same as the technology of embryonic manipulation, and doing cloning work will push gene engineers way up the learning curve toward, say, IQ manipulation. More to the point, the psychology of cloning will break the ice. As the *Economist* noted recently, "a good bet is that attitudes to genetic research will be defined by the first human clone."[52] If she's a "normal" baby, cute in the way that all babies are cute, cute like the first cloned kitty, who was on every front page in early 2002, then public opposition to genetic manipulation may die away. "Supporters of cloning anticipate that the birth of a human clone would quickly erode any public sentiment against it," one critic wrote. "Human cloning is seen by its proponents as a threshold technology that, if allowed, would make it difficult to subsequently draw the line at any of the new human techno-eugenic procedures or applications."[53] On the other hand, stopping human cloning could serve as a firebreak of sorts, what the ethicist Leon Kass has called "a golden opportunity to exercise some control over where biology is taking us. . . . Now may be as good a chance as we will ever have to get our hands on the wheel of the runaway train now headed for a post-human world and to steer it toward a more dignified human future."[54]

So the debate in Congress centered on the question of whether there was a way to allow embryonic stem cell research to proceed without making the production of human clones more likely. Could we have the possible benefits of a technology on this side of the enough point without creating the clones that would clearly colonize Earth 2? A presidential panel headed by Kass recommended one possible way out: instead of a ban on "therapeutic" cloning, a four-year moratorium, during which time regulators would try to work out some scheme to register and inventory individual embryos, prohibit their shipping, and "set a definite time limit and developmental stage beyond which a cloned human embryo may not be grown."[55] They hoped, in other words, to figure out a way to let researchers walk right up to the line without going over it.

By the time they've figured out those guidelines, it's entirely possible they won't even be necessary, because normal medical science, the kind that poses none of these particular philosophical problems, continues to progress—continues to find ways to achieve many of the same goals without raising any red flags. Closing off germline manipulation and temporarily shutting down embryo cloning doesn't mean turning our backs on sick people.

Consider stem cells, for instance. For a long time, biologists thought only the cells of early embryos could do the trick of turning into brain or liver or heart or skin. Adult cells, they reasoned, had already specialized—heart cells could produce only other heart cells. But it turns out they were wrong: the adult body, it's been discovered in the last few years, "contains some rather unspecialized stem cells, which wander around ready to help many sorts of tissues regenerate."[56] Though this is still what one researcher calls a "thin-ice science" that may yet prove a dead end, adult stem cells have apparently been found in blood, in bone marrow, in the brain, in the pancreas, even in fat, "more easily and cheaply available than cells harvested from specially manufactured clones."[57] And they seem to work; in one study, they effected a "permanent reversal" of Type 1 diabetes in mice.[58] Adult stem cells harvested from bone marrow appear to "improve blood flow in otherwise untreatable coronary arteries," a Japanese team reported recently after trials on five patients.[59] Five years ago, a Washington girl named Savanna Jantsch was nearly dead from leukemia; an infusion of stem cells from umbilical-cord blood "built an entirely new blood-cell system" for her. "It's like she's been two different people with different lifestyles and a different existence," said her father.[60] You want miracles? A Canadian lab has demonstrated that nerves can regenerate in the spine when cells from the intestine are transplanted into a severed spinal cord. So far trials have been done only on rats, but every single rat responded to the adult stem cell treatment. If the technique works with humans, only a few centimeters of spinal cord regrowth would be enough to restore arm and hand movements to people whose injuries would otherwise leave them quadriplegic.[61] All that without any existential risk—all that taking place in *this* world.

And stem cells aren't the only game in town, anyway. Our fast-growing knowledge of the human genome opens up the possibility of germline manipulation and somatic therapy—but it also allows researchers to follow a hundred other paths, most of which offer promise without peril. "According to a recent report by the research director for SmithKline Beecham, enough sequencing data are already available to keep his researchers busy for the next twenty years," developing early-detection screening techniques, "rationally designed vaccines," and a host of other useful treatments.[62] Several companies are "actively working on drugs that stimulate the brain and other organs to grow new cells and repair themselves," exactly the same target that stem cell researchers have set. One company is already testing an oral medication on Alzheimer's patients; another has used a similar formula to treat mice with damaged hearts.[63]

As doctors learn more about genetic and molecular biology, they're individualizing treatment. Take cancer, for example. Instead of blindly trying to slay tumors with chemotherapy, then radiation, then some more chemo, "molecular medicine will allow diagnosis of the genetic make-up of the tumor, enabling a targeted treatment with a much-higher chance of succeeding." With the new knowledge of molecular structure, drugs are able to home in on chemical signals that turn a normal cell cancerous—receptors on cell surfaces for growth factors, for instance. Herceptin attacks certain breast cancers; Rituxan goes after non-Hodgkin's lymphomas; Gleevec targets particular kinds of leukemia. "It's the beginning of a sea change—and I am speaking conservatively—in the way we practice cancer medicine," said Dr. Larry Norton, the president of the National Society of Clinical Oncologists. One doctor who'd watched a patient's intestinal cancer—a growth the size of two footballs that had spread to his liver—shrink 80 percent in a matter of days called it "not much short of miraculous."[64]

Other novel approaches have stirred excitement too—so-called anti-angiogenesis drugs that cut off the blood supply to tumors, for instance, and new "aromatase inhibitors," a type of hormone therapy that seems to improve the odds against breast cancer recurrence.[65] These are not all miracle cures, but over time such advances add up to real differences in survival rates. Not long ago, writes the science journalist Natalie Angier, "the word 'cancer' was a verbal anvil, flattening all nuance, sense, and hope. Doctors didn't tell patients; family members didn't tell friends." But no longer; data from the National Cancer Institute show at least 9 million Americans "with a history of malignant disease, many of them in the workplace."[66] One person out of every thirty is a cancer survivor. And all of this with normal medicine, with progress that lies neatly on this side of enough.

When I was a boy, my best friend in the world had cystic fibrosis. We spent each summer with Kathy, one of the happiest and kindest people I've ever known, the sort of person who makes everyone around her happier and kinder. But CF is a cruel disease. Kathy had spent weeks in a coma as a small child, and it had taken an unfixable toll on her coordination. She was frail, her life punctuated by stays in the hospital, her days interrupted by endless "drainage" treatments, her parents rhythmically clapping her back to bring up the mucus her treacherous lungs kept producing. I remember thousands of hands of crazy eights while she breathed an aerosol mist through a face mask; I remember her final summer, when to my later and eternal shame her illness sometimes seemed a burden on *me*, a drag. There were so many things she simply couldn't do. Kathy was two weeks short of her fifteenth birthday when she died; I was sixteen, and not a day's gone by since that I haven't missed her. In the intervening decades, treatment has steadily improved. CF patients routinely live into adulthood now, often long enough to have children of their own. And yet it's still one hell of a tough fate. There's a gene awry, a protein out of balance—and suddenly your body is trying to drown itself.

Why not at least let the germline engineers go to work on the Kathys of the world? They're already trying somatic repairs, injecting healthy cells into existing patients. But why not free them to someday use their techniques to take an embryo that would develop into a little girl with cystic fibrosis and tweak that embryo to remove the CF gene? What harm there, for the 2 percent of people born with CF or other clearly genetic sadnesses?

The harm, as I've tried to show, is not to the patient but to the world in which she lives. As even proponents acknowledge, the line between repair and enhancement is too murky to be meaningful. Soon you're heading toward the world where Kathy's lungs work fine, but where her goodness, her kindness, don't mean what they did. Where someone's souping up her brains or regulating her temper, not just clearing up her mucus.

But it takes a hard person to say that; I'm not sure I'm hard enough. Thankfully, there's one more answer to such tragedy. The technology now exists to allow doctors to screen for diseases like CF. "Preimplantation genetic diagnosis" (PGD) is another technology that walks right up to the enough line, that requires enormous care. But it also clearly allows parents to eliminate the risk of genetic disease without genetic tinkering.

To understand it, think about the kinds of genetic screening we already use. Parents can be tested to find out if they carry a particular disease. If they and their partner each have the gene, they're informed of the odds that their child will as well. If they go ahead and conceive, techniques like amniocentesis allow them to test the fetus in the womb and perhaps opt for an abortion. Such approaches have cut the incidence of Tay-Sachs disease among American Jews by 95 percent in recent years.[67]

But now take it a step further. Say that you are concerned enough about the possibility of genetic disease that you are willing to conceive outside the womb—to have eggs harvested, and mixed with sperm, and then implanted. You'd have to take this step if you wanted germline genetic engineering performed on your embryo, but, as it turns out, you wouldn't even need the engineering to avoid the genetic disease. Instead, each embryo could be screened; only the ones without the trait would be implanted in you and allowed to come to term. It's normal to fertilize more than one egg anyway when using in vitro fertilization—now, you'd simply examine them using our new understanding of what genes lead to what conditions. If four eggs have been retrieved and fertilized in vitro, and if both parents carry a single copy of a recessive gene for a given disorder, then chances are that three of the four embryos tested will turn out to be healthy. If the gene for a given disorder is dominant, and one parent actually exhibits the condition, only a quarter of the embryos will be similarly affected; the rest can be implanted as is, with no need for tinkering. Even in the extremely rare case where both parents have the same dominant-gene disorder, 25 percent of the embryos will be okay.[68]

This technique is not just theoretically possible—it's already been used. As early as 1992, an English baby—Chloe O'Brien—was born in London's Hammersmith Hospital to parents who each carried the CF gene. That meant they had a one-in-four chance of producing a child with cystic fibrosis; they already had a son with the disease. Doctors removed several eggs from Michelle O'Brien's ovaries, fertilized them with the sperm of her husband, Paul, and then stepped back to watch the cells divide in the fertilized eggs. As soon as each had reached eight cells in size, the doctors used a hollow needle to extract one of the cells and test it for CF. Two of the embryos that passed the test were implanted; one developed, and some months later: Chloe. Whose lungs work right. Who will not spend her afternoons being "drained."[69] Who will not, dammit, die at fifteen.

By now, this technique is common enough that only unusual cases make the papers. A Chicago woman who carries the gene that causes a rare form of early-onset Alzheimer's disease, for instance, recently gave birth to a child who had been screened as an embryo to make sure that she would not grow up to become demented at forty. Ethicists applauded the mother for making sure her child would be healthy, but wondered about the morality of bearing a child she herself might have deteriorated too much to recognize by the time first grade rolled around. The case, described in the *Journal of the American Medical Association,* marked the first time Alzheimer's had been weeded out by PGD, but the author gave the real news almost in passing, for the same technique had already been used more than three thousand times to combat diseases like sickle-cell anemia.[70]

As I said earlier, this technology walks right up to the enough line: the same kind of screening that tells you whether your child will have cystic fibrosis can also tell you about plenty of other traits. Like whether that child's a he. According to the essayist Margaret Talbot, writing recently in the *Atlantic Monthly,* one of the country's leading fertility clinics already offers PGD screening for "family balancing"—for allowing people to decide whether they want a boy or a girl. It allows clients, in her words, "to use a medical procedure to select or reject a child on the basis of a characteristic that has nothing to do with life or death, that is not in any sense of the word pathological, that cannot possibly be construed as sparing a child any pain or suffering."[71] It allows rich people, that is, to do just what many in India and China have been doing with sonograms and abortions for a decade or two, though more antiseptically.

And it could set us straight off down the path that leads precisely to the set of horrors I began this book by describing. Lee Silver imagines a woman producing a hundred embryos, and then having each computer-analyzed for its "genetic profile"—to learn, in other words, which one is likely to be the tallest, or the smartest, or the blondest.[72] If embryos could be selected "for the gene variants responsible for the genetic contribution to high IQ," points out Gregory Stock, "the average score of children selected in this way might be nearly 120, well above the average score of 100 found in the general population."[73] To give you an example of the slippery slope possibilities, Stock told the annual meeting of the American Society for Human

Reproduction that parents would "want to weed out children who would turn out to be obese or mentally retarded," as if Down syndrome and a spare tire were basically equivalent conditions. In essence, he explained, "the child will have to pass a test before it is even born."[74]

Even at its worst, this wouldn't be as bad as germline manipulation. At least parents would only be choosing from the possibilities nature presents: from the panoply of choices that these eggs and sperm offer. It's enormous power, but not ultimate power. Abused, though, it would be bad enough. Even the thought that Kathy might have been "weeded out" because she had CF is pretty hard to stomach: as I said, she was among the finest human beings I've ever known. But I didn't have to *be* her. I can understand the appeal of preimplantation screening, just as I can understand the queasiness it causes for disability advocates. The real test will be if we can limit the choices, limit the information, to the relatively short list of life-or-death genetic diseases. "What we need to do," writes Francis Fukuyama, "is not ban the procedure but regulate it, drawing lines to distinguish between legitimate and illegitimate uses."[75] If we can, then this technology will help us stay in a world that still makes sense.

Which is not to say that it will remove every single sadness. Say both parents are sick by virtue of having two copies of a recessive gene for a disease, instead of just being carriers. In that case, all their embryos would manifest the same condition; there'd be nothing to implant. As the Canadian commission charged with studying genetic engineering explained, however, "this would be extremely rare, as the average incidence of a recessive disorder is 1 in 20,000. The random likelihood that two affected individuals would mate is therefore exceedingly small. Moreover, even if they do, if both are healthy and functional enough to achieve pregnancy, the condition affecting them cannot be among the most devastating."[76] Nevertheless, if they wanted a child without that genetic burden, at least before we develop a conventional cure, either they'd need to have germline genetic manipulation, or they'd have to adopt. As would those people whose fertility is so low that even IVF doesn't work for them—people who could have a genetically related child only through cloning. For some of that small number, adoption may represent a defeat, a minor tragedy; the desire to see your biological child can be strong. But not so strong as to warrant the risks that society will take if it moves toward germline genetic engineering. Not so strong that it should push us past the enough point.

But what about helping the poor? If the sick don't require a brave new world to cure them, then maybe the huddled masses do.

In some ways, it's hard to take this proposition seriously. We live in a country that spends five cents per person per day to aid Africans. It's laughable to imagine that before long we'll be, say, building IVF clinics across the continent in order to engineer African embryos with resistance to HIV. Perhaps someday the new "gene vaccines" will make it to the Third World; those are clearly somatic technologies, which all of us will cheer. But "enhancement" isn't headed for the poor of the Third World soon, nor are any of the other technologies we've been discussing. It will be a *very* long time before even the cheapest robot is cheaper than an Indian peasant.

Still, the argument is in the air, if only as a cudgel for the proponents of a high-tech future to beat their opponents with. Consider, for instance, the genetic engineering of food. This book is about people, not genetically modified plants or animals. They raise a separate set of dilemmas, some of which I tried to outline in *The End of Nature.* But genetically modified agriculture is probably the only kind of genetic *anything* that will make its way to the poor world in the near future. And so it nicely illustrates the folly of trying to wave away all sadness with a high-tech wand.

You've probably heard of golden rice, for instance; the biotech companies have spent tens of millions of dollars hyping it. And it sounds like a good idea. A Swiss scientist, Ingo Potrykus, announced a few years ago that he had devised a way to insert a gene from daffodils into rice, so that the grains now had a gene that caused them to express beta-carotene. That is, the yellow in the daffodils was now producing vitamin A in the rice. Since millions of Asians lack sufficient vitamin A, and some of them go blind each year as a result, it sounded like a godsend. "He visualized peasant farmers wading into paddies to set out the tender seedlings and winnowing the grain at harvest time in handwoven baskets. He pictured small children consuming the golden gruel their mothers would make, knowing that it would sharpen their eyesight and strengthen their resistance to infectious disease."[77] Pretty soon the biotechnology industry was running commercials on Western TV, with happy peasant children, beaming rural doctors. By most accounts, airing the first round of commercials cost half as much as producing the rice. "Unless I'm missing something," wrote Michael Pollan in the *New York Times Magazine,* "the aim of this audacious new advertising campaign is to impale people like me—well-off first-worlders dubious about genetically engineered food—on the horns of a moral dilemma. . . . If we don't get over our queasiness about eating genetically modified food, kids in the third world will go blind."[78] Right he was. High-tech boosters quickly charged that environmentalists were keeping the rice from its intended beneficiaries. The right-wing writer Ronald Bailey declared that "a lifesaving grain is being held hostage by anti-science activists."[79] And the campaign worked pretty well; even Bill Joy has said, "I believe we would all agree that golden rice is probably a good thing if developed with proper care."[80]

As usual, though, the world turns out to be more complicated the more closely you look. Gorasin is a village on the edge of the Louhajang River in the district of Tangail in the nation of Bangladesh. It has nothing that we would recognize as a store, no cars, no electric lines, no television. Just small fields, a cow, some chickens, barefoot children, banana palms swaying in the breeze. The call to prayer drifts over the paddy fields from a tiny mosque. "Bangladesh" is our ten-letter synonym for misery, and even in Bangladesh, Gorasin is pretty far from the center of things. Its residents define marginal. And yet, as strongly as any place I'd ever been, it suggested a future for the Third World that didn't run straight through our technotopia—that bypassed it in interesting ways. Gorasin, obviously, was extremely poor. But that wasn't all it was.

In the 1960s, in the heady early days of the so-called Green Revolution, Gorasin's farmers converted to the new high-yield rice strains coming from Western labs. And indeed, their yields grew: "The rice jumped out of the paddy," said one older man, who was bare-chested and wearing a sarong. But the new strains demanded lots of fertilizer—more and more each year, and even so the yields slowly fell. And since the villagers were now planting fields full of identical, "improved" seed instead of dozens of different varieties, they needed lots of pesticides too. Over time, that began to cause trouble, because Bangladesh is as much water as it is land, a soupy river delta where gentle floods are a regular part of life. (It's the floods that make the land so fertile, that allow 130 million people squeezed into a space the size of Wisconsin to feed themselves.) "When we women went to collect water, we would be affected," one villager told me as she stood in a field near her family's hut; she was in her twenties, beautiful, gregarious. "Our skin would absorb the poisons. We would get itchiness, get gastric trouble." A man spoke up: "The cows used to eat the grass and drop dead. And then the villagers would fight each other." Another: "We grew up with a saying: 'Bengalis are made of rice and fish.' Then the fish started catching diseases. We are not scientists, but we made the connection between pesticide and fish death." These people, remember, were not shoppers at Whole Foods who had read some brochures about pesticides. No one had come and told them that they were feeling itchy. They *itched.*

So, a decade ago, when some area farmers began experimenting with a low-tech scheme called "Nayakrishi," or New Agriculture, several residents of Gorasin went to hear a presentation. Before long, they had convinced their neighbors to set up what we could call an organic farming zone—an area the size of a small county where all the peasant farmers, most of them with holdings of no more than an acre or two, pledged not to use pesticides or fertilizer. Instead, they experimented with dozens of different varieties of crops: not only different strains of rice, but sesame seeds and loofah sponge gourds, eggplant, sugarcane, bamboo. Onions, pulses, thirty different leafy greens.

That list is important because many of the things on it provide vitamin A. In fact, it turns out that everywhere the Green Revolution went, it did two things: increase grain yields and increase micronutrient deficiencies. The pesticides protecting the rice poisoned the plants growing on the cultivated margins, so people stopped eating them; and, what with seed and fertilizer and other high-priced inputs, peasants tended to plant every available inch of land to make their money back. All of a sudden millennia-old diets had to change. People were filling their bellies with rice, which was good, but their bellies were otherwise mostly empty. Or, as Dr. Peter Rosset, codirector of the advocacy group Food First, once said, "People do not have vitamin A deficiency because rice contains too little vitamin A, but because their diet has been reduced to rice and almost nothing else."[81] A situation that the golden rice would do little to alleviate: as Michael Pollan noted, "An 11-year-old would have to eat 15 pounds of cooked golden rice a day—quite a bowlful—to satisfy his minimum daily requirements of vitamin A."[82] Furthermore, if you're deficient in vitamin A, you're likely lacking all sorts of other nutrients as well. By some estimates, five times as many humans are short of vitamin B.[83] What you need is not miracles from Monsanto; what you need is a diet rich in local greens. The combination of rice and moringa leaves, for instance, which grow in every country with vitamin A deficiency, provides far better nutrition than golden rice.

Sometimes researchers think peasant farmers, being poor, are therefore stupid, and in need of easy, prepackaged solutions from the magicians of the West. "Transgenic crops are less complicated," explained Wellesley professor Robert Paalberg. "If you have smart seeds, you don't need to have farmers with the same instant upgrade of knowledge. It's actually easier for low-resource, illiterate farmers."[84] But that's simply ignorant. "When we cook the green vegetables, we are aware not to throw out the water," one woman in

Gorasin told me. "Yes," said another. "And we don't like to eat rice only. It tastes better with green vegetables."

This may sound "unscientific." But peasant farmers in places like Bangladesh are starting to gather and trade crop varieties with as much careful attention as any biotech lab worker. At the local Nayakrishi training school in Tangail, farmers had planted twenty-five strains of papaya and 112 varieties of the lovely jackfruit, each catalogued by taste, size, color, season, habitat. Clay pots in a darkened shed contained three hundred varieties of local rice, twenty kinds of bitter gourd, eighty-four types of local beans. "Do you know how much it costs to build a gene bank like the ones where botanists store plant varieties?" asks Farhad Mazhar, a founder of the Nayakrishi campaign. "No scientist can afford to catalogue hundreds of varieties of rice. But farmers are doing it here as part of household activity. Our little seed station has more vegetables than the national gene bank, which spends millions. But we can do it for free."

Mazhar claims that with a combination of traditional and new methods—cover crops, green manures, and so forth—the Nayakrishi farmers can double yields from their croplands. And he may well be right. A number of academic researchers have found real successes using such low-tech agricultural approaches across the developing world. The University of Essex, in what was billed as the largest study ever conducted of sustainable agriculture in the Third World, looked at two hundred projects across Africa, Latin America, and Asia; "on average there was a 93 percent increase in per acre food production with the adoption of techniques such as planting legumes, using integrated pest management, and introducing locally appropriate crop varieties and animal breeds." They described Honduran fields turned from hardpan into "thick, dark, spongy soil almost a meter deep" through the use of compost.[85] An experiment across 100,000 acres of Chinese paddy land showed that farmers could nearly eliminate the worst crop blights and double yields without using any chemicals, simply by interplanting different varieties in their fields. Hans Herren, who won the World Food Prize in

1995 after he helped figure out how to use a parasitic wasp to wipe out the biggest threat to the African cassava crop, says such work is essential—but he doubts he could even obtain funding for it today. "Today, all funds go into biotechnology and genetic engineering. Biological pest control is not as sexy."[86]

It's not that poor people have no need of technology. Solar panels can't sprout fast enough; the first ones had recently been installed near Gorasin, allowing evening literacy classes for farmers. A vaccine against malaria or dengue would be a great boon. But the fact that one technology improves lives says nothing about the next technology. All that "transgenic crops have done so far is tie the farmer to specific chemicals and a specific company," says Tewolde Egziabher, Ethiopia's environment chief.[87] Most researchers acknowledge that figuring out ways to alleviate poverty—with land reform, for instance—would do more to alleviate hunger, simply by enabling people to afford food. But genetic engineering tends to lead in the opposite direction, helping prosperous larger-scale farmers the most.

The men and women of Gorasin, "illiterate" and "low-resource" though they undeniably were, seemed to understand all this almost intuitively. I was in their village the day an agricultural scientist first came to talk about golden rice. As she described the new rice, farmers kept jumping up to deliver little lectures of their own. They knew, immediately, that it would mean pesticides and fertilizer, and they knew they wanted none of it. We sat in a courtyard as the evening came on, in a semicircle formed by the village's small huts. Thirty-five or forty people were on hand, pretty much the whole population. Two babies were using a grapefruit as a ball, which every person in the village would roll back to them with a smile when it came their way. One of the older men, Akkas Ali, mentioned that he had written a hundred songs praising organic agriculture, tunes he and the other men often sang on market day at the local bazaars in an attempt to convert other farmers. "Food from Nayakrishi is so much better," he sang in a reedy Bengali, as the rest of the village clapped rhythmically behind him. "No longer do I eat the poisons. Why should I eat that life-destroying stuff? If you use organic fertilizer, the Almighty will be behind you. And you'll be having no more gastric problems."

It's easy to romanticize happy peasants singing in the setting sun. I left that night for home. They woke up in the same place, still too poor. Though Bangladeshi women have dramatically lowered their fertility, there's still enough demographic momentum to perhaps double the country's population again, an almost unimaginable crush of people. Perhaps there's no way out but ahead through the high-tech future.

But it's easy to romanticize that high-tech future too, as Monsanto does every time the company helps fund one of those smiling-brown-child golden rice commercials. People in Gorasin have had more experience with the enough point than most of us—they're the ones who waded through waters filled with pesticides, the ones who watched their fish catches dwindle. At least for the moment, it's clear to them that whatever they may lack, it's unlikely to be supplied by the technotopians.

So poverty and illness may not require the highest tech; maybe they can be dealt with in the world we currently inhabit, with the kind of small and steady scientific and cultural progress we're used to. But there's one "problem" that clearly *can't* be dealt with this side of the enough point. It's the chief puzzle of human existence: the fact that we grow up, and then grow old, and then cease to be. Our sense that we're incomplete, in need of improvement, doesn't really come from the scandal of our multipurpose mouths, or the traumas of reproduction or elimination. Freud was right when he said that we were repressing something—but, as the psychologist Ernest Becker pointed out, it wasn't sex. "Today we realize that all the talk about blood and excrement, sex and guilt, is true not because of urges to patricide and incest and fears of actual physical castration, but because all these things reflect man's horror of his own basic animal condition. Consciousness of death is the primary repression."[88]

At the moment, as in all of human history, we can deal with this fear in three main ways: denial, stoicism, or the kind of hope represented by our various religious faiths. But now, all of a sudden, the door that leads into a fourth room has swung ajar. It opens onto a laboratory, where people have begun to speak in serious tones about—about curing death.

Surely they are charlatans or kooks? Some of them, yes. Not long ago I drove to southern Quebec to interview the now world-famous Rael. A French former sportswriter and small-time race-car driver, he began his own spiritual movement after his reported abduction by aliens in the early 1970s. For a long time he concentrated on raising funds for the embassy that the aliens wished him to build before they would return. But he hit the big time in the late 1990s when, post-Dolly, he spent a thousand dollars to open a post office box in the Bahamas, a front for a nonexistent cloning operation. "I had no actual intention of having a lab there," he said with a laugh. "The goal was to have a big media coup for about nothing. We got twenty million dollars in publicity for a thousand dollars. We were laughing our eyes out about it."

Once he'd seen the result, it's easy to imagine that Rael almost wished he'd never started in on the spaceship stuff. Suddenly his white jumpsuit (it looks as if someone has thrown nanostrength bleach in the *Enterprise* washing machine) and his special topknot haircut, which serves as an antenna, seem a tad retro. Whereas cloning . . . Before too long, Rael had turned his Bahamas mail drop into something like a real cloning effort, with actual scientists and plenty of young female followers willing to serve as surrogate mothers. At UFOLand, his Canadian headquarters, there's a big model of the double helix mounted right next to the Elohim spacecraft. Now his voice rises with excitement when he starts talking about all the new technologies. Here is Rael on human genetic engineering: "Let's say you have a family of artists, and they want to pass it on to their child. But maybe he will want to study mathematics. But if he has a genetic modification to make him more sensitive to the arts instead—well, then, everybody is happy." On nanotechnology: "Soon you will have a little kind of microwave. Not with a button, because a button is a thing of the past, but with a microphone. You will say, 'Give me a salad. A lettuce.' Soon a perfectly healthy salad. Then, you want a leg of chicken. Without the chicken. Without the farming." On linking your brain to a hard drive: "With a higher-level consciousness, to still have to pee, to shit, to have sweat? No. I will be cleaner as a computer. I will have to defrag sometimes, but that's it." (Appropriate chuckles from all acolytes in attendance.)

In the end, though, what really attracts Rael to this technofuture is the hope that it will let us live forever. He cared less about Eve than Methuselah. His latest book, *Yes to Human Cloning,* is subtitled *Eternal Life Thanks to Science.* "Right now, cloning is for people who aren't having a baby. That's not so interesting. We support it because it's a first step. But the top of the steps is eternal life."[89]

Rael is a cult leader, by far the most genial and frank cult leader I've ever interviewed. He is, I think, a nut. But you couldn't prove his nuttiness by his obsession with immortality. Compare his words with those of, say, Michael West, the CEO of Advanced Cell Technology, the leading firm in the human genetic engineering business. "All I think about, all day long, every day, is human mortality and our own aging," says West.[90] "In a thought experiment, we can imagine making body components one-by-one, each made young by cloning. Then our body would be made young again segmentally, like an antique car is restored by exchanging failing components. . . . I can take a cell from a 100-year-old-person and make a brand-spanking-new dimpled babe—doesn't that in itself tell us something about the immortal substratum of life?"[91]

Michael West is not, as the world reckons it, a nut. He's the man who first cloned embryos and grew them to the six-cell stage. He's been quoted deferentially in a thousand news stories; when he testifies before Congress he's the prince of sweet reason, interested only in curing genetic disease. He complained bitterly when the House wanted Rael to come talk. "This is absurd. It's a circus. Why is Congress debating this by talking to someone who says he flies around in flying saucers?"[92]

In fact, however, when West lets down his hair in interviews with obscure tech publications ("*Ubiquity,* the online magazine of the

Ultranet"), it turns out he's interested in precisely the same kinds of immortalism as Captain UFO. "A self-described political conservative, he was in his early years a creationist, and he trained as a paleontologist with the goal of proving the Bible's account of God's design," writes the journalist Joannie Fischer. "But as he studied the fossil record, instead of finding God's divine plan, he found an endless account of disease and suffering. Out of that bleak vision he developed a new spiritual fervor. 'If God is about love and life,' he says now, 'then we should do everything we can to end suffering and death.'" After earning his Ph.D. in biology, West reincorporated his late father's truck-leasing business as a biotech firm called Geron, Greek for "old man." After seven years, the company eventually succeeded in cloning telomerase, an enzyme that keeps cells from aging.[93] But by then the company's board members were courting Wall Street, and they began to worry that their chief executive's obsessive discussion of "curing" old age sounded bizarre to investors. He had moved on to stem cells anyhow, so in 1998 he quit and went to work for Advanced Cell Technology, a Worcester, Massachusetts, firm that cloned cows for farmers.

His backing came from, among others, Miller Quarles, an octogenarian geophysicist who'd made a fortune finding oil and was only too happy to spend it finding the gusher of youth. "I'm not investing in him to get rich," Quarles explained. "I want to live another hundred years." Along with helping to bankroll West, Quarles, who swallows fifty-five vitamin pills a day, founded the Cure Old Age Disease Society. Says West, "Miller is not afraid to look death in the face and say 'let's not die anymore.'"[94] As for West, he has no problems with self-esteem. "I once saw a T-shirt that had a picture of Einstein, under the face it read, 'If he was so God damned smart, why is he dead?' The point is, aging is a disease that is killing everyone."[95]

A *New York Times* reporter who profiled West reported that "his entrepreneurial interest in the biology of aging derives from an obsessive, almost morbid fascination with death."[96] In this, West is not alone. The dream of immortality is shared by many of the people driving these new technologies forward; it is the secret force pushing them to push the envelope. We don't need these technologies to treat illness; we may not require them to prevent hunger in the Third World. But we do need them if we're going to conquer death.

Consider just a few of the people we've already met. Gregory Stock, for instance, the advocate of human genetic engineering: "A sad irony of life is that brutal decay is the fate in store for each of us. . . . Although preimplantation genetic diagnosis may enable us to protect our children from various age-related diseases, it will at best only modestly defer the reckoning with Father Time. In light of our yearnings for immortality, the underlying biology of aging may well be the first germline intervention to truly tempt us."[97] Or Damien Broderick, author of *The Spike:* "We're stuck at the moment with death's pain, loss, and grief. . . . But in the longest term of the history of intelligent life in the universe, it will surely be the case that the routine and inevitable death of conscious beings was a temporary error, quickly corrected."[98] Or Robert Freitas, one of the chief nanotech theoreticians, and author of technical papers explaining how the "gray goo" problem might be overcome: "Until nanomedicine becomes available, nobody can have any reasonable expectation that they will live long enough to pursue their longest-term and most interesting goals to fruition. Thus the development of nanomedicine looms like a tollgate, separating each one of us from the rest of our lives. Nanomedicine is the serious futurist's *sine qua non.* It is our gateway to all that will be, and all that we can become. That's why I personally regard it as my top priority."[99]

These aren't the first people to be unhinged by death, of course. A reasonable working definition of a human being is any creature unhinged at the prospect of dying. "*This* is what is creaturely about man, *this* is the repression on which culture is built," insists Ernest Becker.[100] That culture covers everything from the pyramids of Egypt to the "Left Behind" novels now crowding the top of the best-seller charts. Death's overpowering reality drives some to embrace various creeds, or to mummify bodies, or to jog. Some of us try to achieve glory so our names will be "kept alive"; others to reduce ourselves until we merge with some ultimate reality. For every person eating and drinking and making merry in the face of death, another is abstaining, hovering around the vitamin aisle at the health food store. Death is us.

And coping with death may well have gotten harder in the modern world, complicated our longings. Science made it an act of faith against reason to hold on to our religious hopes. It was Pascal who wrote as early as 1670: "Let us imagine a number of men in chains and all condemned to death, where some are killed each day in the sight of others, and those who remain see their own fate in that of their fellows and wait their turn. . . . This is the picture of the condition of man." Scientists tended to pride themselves on their ability to face the cold, hard facts of life, to come to terms with the nothingness they knew would follow, and to hold in some contempt the "fairy tales" of the pious. (No small irony, then, that some among them now turn out to be so afraid of dying that they're willing to turn the earth upside down to prevent it.)

But it's not just science that's made death scarier. Our lives have, by and large, become easy and comfortable; we no longer feel so worn down by living that we can imagine death as any sort of relief. Instead, the kings and queens of entitlement, we toss about like the pharaohs for some alternative. "Baby boomers are bummed," jeered one technotopian. "For all the technology, perpetual prosperity, and material gains they enjoy, they are approaching the time when they are going to keel over like their horse-riding grandparents." "And there is nothing we can do about it. So far."[101]

Here's the problem. On the one hand, we're clearly living longer all the time. A child born today in America can expect to live a little more than seventy-five years; on average, twenty-eight years longer than a child born in 1900. Those who make it to sixty-five can expect another seventeen years of life, six more years than awaited a sixty-five-year-old in 1900.[102] And Westerners aren't the only ones living longer; though AIDS has played havoc with African lives and African demography, anyone who reaches the age of sixty-five in, say, Tanzania is likely to outlive his or her counterparts in the rich world.[103]

Most of the gains, in every corner of the world, have come from better public health. In 1900, Americans succumbed most frequently to tuberculosis, then to pneumonia, and then to diarrhea. Heart disease ranked fourth, and cancer eighth. Now, thanks to sewers and refrigerators and antibiotics, we live long enough that those latter two scourges are our leading killers.[104] As we have seen, medical science continues to improve, and life expectancy continues to increase. A study of women in New Zealand, Norway, and Japan found that their longevity was increasing by three months per year, with no sign of stopping soon.[105]

But even as our life expectancy increases, there's no sign at all of immortality. Indeed, it becomes steadily clearer that our possible life span is fixed. Exercising vigorously your whole life may add about two years to that life, but that's it (and, according to one Stanford University study, most of that two years will have been spent exercising).[106] We still lose strength and stamina as we age. Our vision deteriorates, hair grows in our ears and nostrils, we lose bone mass. "As we age, thousands of changes occur in all of our organs and tissues, in the individual cells that compose them, and even in the cement that holds our cells together. Age changes occur even in the individual molecules of which our cells are composed and in the products that our cells produce."[107] We wind down.

It took researchers most of the twentieth century to unravel how and why this happens, and to truly document its inevitability. And to do it, they had to defeat an earlier generation of the technofaithful. In 1912, Alexis Carrel, a French doctor working at the Rockefeller Institute in New York, announced that he had succeeded in culturing tissue from the heart of a chick embryo, and that instead of the normal three to fifteen days of life outside the body, his cells were still going strong after three months. He titled his paper "On the Permanent Life of Tissues Outside the Organism," and wrote that "it is even conceivable that the length of the life of a tissue outside of the organism could greatly exceed its duration in the body because elemental death might be postponed *indefinitely* by proper artificial nutrition."[108] In other words, Carrel said he'd found something that could live forever. And for the next thirty-four years, until they were "retired," his tissue cultures seemed to do just that. Reporters were allowed to peer through glass portholes into the lab, which was painted black to reduce the chance of microbial contamination. Inside, writes the historian Carol Kahn, "technicians capped and gowned in black, like the votaries of some medieval religious sect, tended their 'immortal' cells." It's true that other scientists had trouble repeating Carrel's feat, but there was always a reason—"You didn't prepare the culture media properly, or the glassware wasn't clean enough, or the phases of the moon were wrong, or the technician was drunk the night before," recalls Leonard Hayflick, the man who went on to disprove Carrel's claims, and with them the notion of immortality.[109]

"Our discovery occurred quite unexpectedly, as is often the case in scientific research," Hayflick wrote years later. It was 1959, and "we were growing cells from human embryonic tissue obtained from legal abortions sent to us from Sweden, trying to find viruses that might be implicated as the cause of human cancers." As they cultured the human embryo samples, the researchers noted that the cells only grew and divided a certain number of times before dying. "The dogma that I was taught held that, although cells are inherently immortal, they will frequently die unless extraordinary precautions are taken to ensure that the growth conditions are perfect," said Hayflick, and so he wasn't surprised to see them dying off. He was surprised, however, to notice that they died in order—that is, the oldest cultures died, the younger ones thrived. "If there was a mistake in our technique, then *all* the cultures should have been dying at the same time, not just the oldest ones." On closer investigation, he found that each dying culture had divided about fifty times over an eight-month period. Younger cultures, which had divided less than fifty times, thrived—but past fifty divisions, death was inevitable.[110]

It took years for Hayflick and his colleagues to kill off the dogma of immortality; their paper was at first rejected and then challenged. But eventually the evidence was overwhelming, and other researchers began to discover the mechanisms that governed this "cell clock." Chromosomes end in a region called a telomere, which consists of "a number of repeating subunits"; Hayflick compares them to beads on a string. With each cell division, one of the subunits disappears, until all the beads are gone. Only cancer cells can keep dividing forever, because they produce an enzyme called telomerase, which makes more telomeres. In every other case, the telomere is like an hourglass, and the sand runs one way.[111] "It is as if the tips of the plastic ends of your shoelaces get snipped off each time you take your shoes off," wrote Gregory Paul and Earl Cox. "Eventually the tips will be gone and the string will start un-raveling."[112]

The so-called Hayflick limit is one of the great discoveries of science, as profound in its way as the limits imposed by gravity or by entropy. It explains why human beings don't live past the age of 115 or 120: their cells have simply run out of steam and can't keep dividing. People claim, occasionally, to have found a 140-year-old peasant living on yogurt in some Balkan valley, but in fact, "few, if any, birth records for people who claim to be older than 115 can be unambiguously authenticated."[113] So even if we completely eliminated heart disease, we would add only four years to the average life expectancy; getting rid of cancer would add only two.[114] (Says Hayflick, "Even if we became able to prevent *all* causes of death now appearing on death certificates, the resulting increase in life expectation would not come close to the twenty-five-year increase that has occurred since the turn of the century.")[115] This means that life expectancy is not going to change radically. "Generally, people have genes that should be getting them to their mid-80s," says Dr. Thomas Perls, the director of the New England Centenarian Study. "In our country we live about ten years less than that on average because of terrible health habits." (Even Okinawans, whose fish and soy diet and abstemious ways make them the longest-lived people on earth, last only five years longer than we do.)[116] The best we could possibly hope for is what demographers call a rectangular curve, meaning in this case that everyone would "slip away peacefully during a narrow span of years, say between age 110 and age 115."[117]

So we can expand our working definition of a human being: we're the ones who worry about dying and who max out at 115. That's who we are.

But maybe, just maybe, not forever. The Hayflick limit had a brief reign, perhaps a decade, as the last word. Now, on the margins and increasingly from the center, researchers are rising up to suggest ways it might be circumvented. In place of the tools of yore (golden bathtubs, sex with virgins, transplants of goat glands), they want to use the new techniques described in this book.[118] As early as the mid-1980s, for instance, scientists had discovered that the rate at which DNA repaired itself correlated closely with longevity; preliminary research also led them to believe that you'd need to make changes in relatively few—perhaps a hundred—genes to double life spans.[119] In more recent years, they've begun to learn which those genes might be. An Icelandic study published in 2002, for example, found a stretch of DNA shared by many of the island's nonagenarians, a stretch of DNA that produced a particular protein that "is helping people live to ripe old ages."[120] The news did not come as a complete shock: in 1990, scientists had managed to increase the life spans of roundworms by 50 percent, simply by manipulating several genes. A gene known as SIR2, which had previously increased the longevity of yeast cells, let the worms live three weeks instead of two.[121] Other researchers had tinkered with different genes in other species of worms, allowing one strain of nematodes to live seven times longer than normal. Meanwhile, in the late 1990s, scientists working for Michael West, then at Geron, announced that they had engineered normal, noncancerous cells to produce telomerase, which kept them alive for two hundred divisions instead of Hayflick's fifty.[122]

Some of the genetic researchers were especially interested in dieting. Just as most Americans were giving up on lo-cal and heading for the steakhouse, scientists were finding that, across a wide range of species, "caloric restriction" could powerfully increase longevity. Rats, guppies, water fleas, yeast, spiders, Labrador retrievers, and monkeys all lived longer if they ate 30 percent fewer calories; researchers had one rhesus monkey in a cage at the

National Institutes of Health who was thirty-eight years old, the human equivalent of 114.[123] Nobody knew for sure whether caloric restriction would work in human beings, but one man, Roy Walford, was bent on finding out. A gerontologist who heard a radio broadcast of *Faustus* at the age of eleven and decided that he, too, would extend his life, but without surrendering his soul, Walford found that he could double or triple the life spans of some fish by underfeeding them.[124] So he began fasting two days a week, and eating lightly in between. At seventy-eight, he's still going strong, as are a growing number of followers. In 2001 the *Wall Street Journal* found a Pennsylvania technology entrepreneur who built himself a basement sprout farm. "If you like arugula, you'd really like arugula sprouts," says Dean Pomerleau, who is five feet, eight inches tall and weighs 127 pounds.[125] But no one's betting that Americans will decide to give up food (for the record, eunuchs also live considerably longer than average).[126] Instead, scientists think that caloric restriction may turn on and off certain genes—and if they can figure out which ones, then an engineered human might be able to have his cake and eat it for two or three centuries.

Gregory Stock, the UCLA researcher who convened the meeting of gung-ho germline engineers, recently helped assemble a similar panel of researchers on aging. With germline intervention, many of them said, humans might soon be able to live twice as long as they do now. Dr. Michael Rose, a biologist at the University of California at Irvine, told Gina Kolata of the *New York Times* about his work with flies: "I've created postponed aging with my own hands," he said. "I know what it feels like to see one organism on its last legs and another organism that is the same age doing fine." Dr. Cynthia Kenyon, a professor of biochemistry at the University of San Francisco, said she'd done the same thing with nematodes. It is, said the forty-five-year-old Dr. Kenyon, as if someone who looks as she does were actually ninety. "'Just imagine it, I'm 90,'" she said, challenging the scientists who turned to stare at her, a tall, vibrant, honey-blond professor. According to Kolata, "most of the scientists at the meeting said the question no longer is 'Will it happen?' but rather 'When?' 'There's nothing bigger,' said one researcher. 'If we could do it, there's nothing bigger. It's the big enchilada.' "[127]

Or, rather, the first bite at the enchilada. Once you're 150—well, why stop there? Genetic alterations might not take you much further, but cloning could conceivably produce new organs, new cells, to put in place of the old; that's what Michael West meant in using the metaphor of a classic car rebuilt from all new parts. "You could take a person of any age—a hundred and twenty years old—and take a skin cell from them and give them back their own cells that are young!" he exults. "Cells of any kind, with any kind of genetic modification! That's such an incredible gift to mankind!"[128]

And where biology leaves off, hardware could conceivably take over. Nanomedicine theorists have, for instance, look forward to the day when medical bots patrol the body, "armed with a complete knowledge of a person's DNA," ready to "dispatch any foreign invaders."[129] One nanotech enthusiast reports that he has "drawn up intricate plans for a system of nanobots that would make blood circulation obsolete. Tiny robot tankers containing oxygen, nutrients, and wastes would replace all the functions of the circulatory system. The heart, that great clunky pump, would no longer beat." Asked if he'd miss the *ba-bump* of a heart, the first sound that any of us hears, the nanodesigner said no: "The noise in my ears keeps me up when I try to go to sleep."[130]

Some of these technologies are clearly closer to bearing fruit than others; the real fear of the futurists now alive is that they won't live to see the great day, that they'll be tackled by the reaper on the five-yard line. Like Moses, they will have visited the mountaintop and seen the Promised Land, but they won't reach it themselves. Which is why, at a Foresight Institute convention or a transhumanist gathering, so many of those present are wearing silver bracelets detailing how they are to be frozen in the event of their demise. The idea of freezing—"cryogenics"—first emerged after the war, when Robert Ettinger wrote a tale for the March 1948 issue of *Startling Stories* about a millionaire who had himself frozen and placed in a vault—"the first of men to die a qualified death."[131] But Ettinger was not content with fiction. In 1962 he self-published a book, *The Prospect of Immortality,* that eventually went through nine editions and fleshed out his vision of cryogenics as "a bridge to an anticipated Golden Age, when we shall be reanimated to become supermen with indefinite life spans."[132]

In its early days, cryogenics was more or less a joke. Reports that Walt Disney had been iced turned out to be urban legend. The first actual freezee was James H. Bedford, a California psychology professor, who was chilled in January 1967; at one point in their work, the "technicians" had to open his coffin in the middle of a public park and throw in more dry ice.[133] Ettinger himself was pretty vague about technique, once suggesting that perhaps bodies should be "shipped to Siberia for natural cold storage."[134] But slowly the field gained a kind of fringe credibility. Companies emerged, Alcor Corporation chief among them, to handle the corpses. When California tried to prosecute the firm for freezing an eighty-three-year-old lady, statements arrived from scientists at Harvard, Columbia, and Johns Hopkins that called the procedure "reasonable." Eric Drexler wrote the court that "future medicine will one day be able to build cells, tissues, and organs, and to repair damaged tissues. These sorts of advances in technology will enable patients to return to complete health from conditions that have traditionally been regarded as nonliving and beyond hope, i.e., dead." Hans Moravec sent his statement from Carnegie Mellon, where he headed the Mobile Robot laboratory: "It requires only a moderately liberal extrapolation of present technical trends to admit the future possibility of reversing the effects of particular diseases, of aging, and of death, as currently defined."[135] Alcor was acquitted; the old lady remains frozen, somewhere in the general vicinity of Ted Williams.

Scientists continue to work in the field—Alcor reports that it has managed to revive dogs frozen for four hours with "no measurable damage"—and if you want to sign up, their busy Web site includes an on-line application and outlines the fee structure ($120,000 for whole-body suspension, $50,000 for just your head). "Signing up for cryonic suspension is mostly a matter of paperwork and obtaining insurance. . . . When your arrangements are finally complete, and you put your ID bracelet on your wrist, you will know that you have raised a shield against the worst that could come."[136] So far there are only a hundred or so bodies in the fridge, but another seven hundred or so are wearing the bracelets, and enthusiasts predict that one good scientific breakthrough (reviving a mouse that had been frozen for two weeks, say) could hand them half a million clients a year.[137]

In the meantime, they're doing their best to convince people that living forever wouldn't be weird or gross or boring. James Halperin, in his novel *The First Immortal,* imagines waking up to meet his great-grandson, a nanoscientist who has "used trillions of tiny machines to put all your molecules back in the right place." He goes on to have some excellent sex with "the most erotic looking woman he'd seen in his lives—both of them," and also to clone his ex-wife so that his son can raise her till she's old enough to remarry him. When he's finally reunited with this granddaughter/wife, he's relieved to find that she kisses exactly the same way he remembered from his first life: "same suction, same moistness, same pressure." In forty-seven weeks they make love 584 times.[138] In case that doesn't quite sell you, Ettinger explains that endless life will make us better people and "provide a panacea, particularly in international relations, because it provides *time* for the solution of problems."[139] Also, the economy would grow faster: Gregory Stock, citing a Yale university study, estimates that "half the increase in the standard of living in the United States during the past century is due to the rise in longevity,"[140] and Damien Broderick points out that "the economic tragedy of aging and death is that so much human effort and resource is expended in education and training that's inevitably lost within a couple of decades."[141] Immortality as a way to amortize tuition!

Objecting even slightly to immortality is a little like arguing against ice cream—eternal life has only been humanity's great dream since the moment we became conscious. And yet we've never had to deal with the possibility that we might actually be able to bring some version of it into being.

Living forever raises a few practical problems, of course. For instance, Robert Ettinger has outlived and frozen two wives, so whom is he married to when they all wake up? ("The rich have their problems and the poor have their problems, but the rich have a better class of problems," he says. "If we are all revived, I will consider that a very high-class problem.")[142] And isn't the world going to get a little crowded? (Even if every woman had just two children over her endless life span, that would no longer represent "replacement fertility" because the children's parents aren't going anywhere—as one futurist suggests, "even with cheap space travel and abundant resources in the asteroid belt," population will "need eventually to be modified by curtailing the number of children born.")[143] And how on earth do you ever get rid of the folks in power? ("President for Life" would take on an even more ominous sound.)

But the real dilemmas go deeper, straight into the realm of meaning. Some of the futurists see these technologies as just one more step along the path of progress. "Defeating death and planning rejuvenation are goals no more absurd than finding remedies for shortsightedness and asthma," writes Damien Broderick.[144] But this is nonsense. So far we are "mortals," literally defined by the fact that we perish. If we don't die, then our lives as we have known them since we climbed down from the trees will no longer make sense; we will be fundamentally different creatures. "The new immortals," in the words of ethicist Leon Kass, "would not be like us at all."[145] A new species, *Homo permanens.* It will be by far the biggest change we've ever faced. But will it be a change for the better?

For the technotopians, as usual, the default assumption is that more is better. If living 80 years is good, then living 160 is ipso facto twice as nice, and living 800 is an order of magnitude better. "If life is worth living, it should not come to an end," writes Michael Perry. "Therefore one ought to be immortal."[146] But occasionally they do bother to argue the point. With longer lives, they say, we'd be able to relax. Freed from the specter staring patiently over our shoulder, we'd have both time and reason to build a better world. "With an unlimited future to redress the balance," writes Robert Ettinger, "everyone can put up with temporary burdens and inequities patiently, and negotiate in good will." People would stop polluting, knowing that they'd pass this way more than once. They'd treat their fellows better, "because there are no more strangers, but only neighbors whom I will have to look in the face again and again."[147] At the moment, insists Broderick, if anyone takes up the piano or studies a new language in middle age, it's "derided as comic evidence of 'mid-life crisis.'" With a life span limited to a single century, "a quarter devoted to learning the basics of being a human and another quarter, or even more, lost in failing health, it's little wonder that we constrict our horizons." But not in the world to come, when we will fill our days with arts and science and "altogether new means of expression" simply for their "own soul-filling sake."[148]

Well, maybe. At the moment, though, I would wager that the prospect of death drives more achievement than it hinders. It "concentrates [the] mind wonderfully," as Samuel Johnson once remarked; perhaps you, too, are one of those people for whom the aptly named "deadline" is a requirement. Perhaps endless life would feel like government work, so safe and secure that initiative is dulled.

But for the most part this discussion is simply silly. Who cares, in the end, how much we all "accomplish"? The question is, who will we *be*? We are, right now, the animal that knows it will die. The one who has eaten of the apple and is by that knowledge changed. Our understanding that we will die is in some powerful way the essence of who we are—and who we are is a paradox. "Man has a symbolic identity that brings him sharply out of nature," writes Ernest Becker. "He is a creator with a mind that soars out to speculate about atoms and infinity. . . . This immense expansion, this dexterity, this ethereality, this self-consciousness gives to man literally the status of a small god." And yet at the very same moment he is food for worms. Ashes to ashes, dust to dust. "He is dual, up in the stars and yet housed in a heart-pumping, breath-gasping body."[149] The other animals needn't deal with this; neither will the robot-men that spin off into the future. In some ways that will be a great blessing, for the prospect of death can be terrifying.

Without it, however, consciousness would carry little meaning; it would have nothing to rub against, it would spin like a tire on ice. We would be disconnected from the body; even if we still had a container, even if we retained the sound of the beating heart out of some sentimental tic, the body would be much more like a car than a carcass. Just something to carry our brain around in.

Without mortality, no *time.* All moments would be equal; the deep, sad, human wisdom of Ecclesiastes would vanish. If for everything there is an endless season, then there is also no right season. No time to be born, nor to mourn, nor to rejoice, nor to die. "Anytime" is not the same as time that matters. The future stretches before you, endlessly flat. The poet Robert Pack, in a stunning small book of essays called *Affirming Limits,* writes that "one endures the sorrows of the ephemeral because they intensify and brighten the momentary awareness of *being.*"[150] If time stretched out forever before you, then you'd never need to really choose; your satisfactions would be the satisfactions of a slow-motion orgy, not of a love.

Nor would you ever have to sacrifice. Without the possibility of death, heroism would disappear—and heroism has always been one of the deeply human callings. ("If a man hasn't discovered something he will die for," said Martin Luther King, Jr., "then he isn't fit to live.")[151] But by heroism I do not mainly mean the courage of the small band at Thermopylae, or Gandhi on the salt flats, or John Bull against the Blitz, or Abraham Zelmanowitz, the man who would not leave his wheelchair-bound friend in the chaos of the World Trade Center. I mean, chiefly, the profound daily heroism of bringing up a child in the full knowledge that he or she will supplant you. In an endless future, there might still be some beings called children, with whom you originally had some financial or tangential biological connection, and who perpetually trailed twenty years behind you.

But you would never pass on your life. Your "child" would be just one more figure in the sea of figures, owing you little and owed nothing in return. (Literally nothing: everyone could be issued one of those bumper stickers that says "I'm Spending My Kid's Inheritance.")

The immortalists imagine that if one bite of the apple gave us consciousness, another bite or two might take away the pain that came with that consciousness—the knowledge that our lives do not go on forever. But it is at least as likely that the next bite will erase meaning instead: that meaning and pain, meaning and transience are inextricably intertwined. That all the harmonies that make human life wonderful and special depend on the approximate shape of a human life. That they can survive the gradual lengthening of longevity we have seen so far, but not the destruction of the Hayflick limit. Immortality, or something like it, is not just *more,* whether or not more is a good thing. It's *different,* completely different.

These reflections are by their very nature preliminary. No generation of writers or thinkers has ever before faced the actual *possibility* of immortality. Always before, it's been a symbolic or religious or figurative prospect, not something to be accomplished with gene regulation or nanotechnological wizardry or silicon-flesh connections. There's been no time to let the idea gestate in our various cultures. But it fills me with the blackest foreboding. It would represent, finally, the ultimate and irrevocable divorce between ourselves and everything else.

The divorce, first of all, between us and the rest of creation. This split is already deep, but it's not total. When the phone call came that my father had been diagnosed with terminal brain cancer, I headed first for my wife, and then, after a hard hug, for the woods. It was only there, amid saplings sprouting from decaying logs, and leaves turning into soil, and summer turning into autumn, that the news made some small sense. Everything goes and comes back and goes again. My friend David Abram, a great scholar of the animist traditions, said once that "in any culture awake to the sensuous world, death is not a big problem, since it's obvious that nothing really vanishes: you just transform back into the soil and the wind and the chattering leaves." Indeed, a kind of existential eternity envelops anyone really awake to the organic entanglement of their own life in the wider life of the land. In such a world, dying is simply one more transaction in an endless gift economy, just as animals and plants have given of themselves to nourish you.

But even in the mostly human world where we mostly now reside, immortality would mean divorce: divorce from every other human being. It would be the final triumph of our individual liberations. Instead of a world where we owe attention to our elders and care to our juniors, we would owe nothing to anyone. I earlier described Arthur C. Clarke's novel *The City and the Stars,* about a man who escapes from the eternal city of Diaspar into a village of real people, people who die and are born. In the village are the first children he has ever seen: he "watched them with wondering disbelief—and with another sensation which tugged at his heart but which he could not yet identify. No other sight could have brought home to him so vividly his remoteness from the world he knew. Diaspar had paid, and paid in full, the price of immortality."[152]

Could such a thing happen here—could such selfishness overtake us and cut us off from all that bears real meaning? Here's Michael West again, the man at the very front of this charge into the future. An interviewer asks him whether immortality won't lead to overpopulation. It's true, he replies, that "an elimination or slowing of human mortality would aggravate the problem," but "why put the burden on people now living, people enjoying the process of breathing, people loving and being loved. *The answer is clearly to limit new entrants to the human race, not to promote the death of those enjoying the gift of life today* [italics mine]."[153]

Now, now, today. He and his colleagues want to stop time. But you *can't* "enjoy the gift of life" forever. Maybe with these new tools you can *live* forever, but the joy of it—the meaning of it—will melt away like ice cream on an August afternoon. It is true that nothing short of these new technologies will make us immortal, but immortality is a fool's goal. Living must be enough for us, not living forever.

CHAPTER FOUR

Is Enough Possible?

Can we, even if we want to, actually rein in these technologies? Can the opposition to them ever be more than academic?

If the answer is no, then all the arguments about meaning and risk are so much fluff, and the idea of enough is a cloud castle. This is precisely what the technological visionaries insist, in the strongest possible language.

"The world is changing, and our humanity within our world is changing with it. The forces of change are irresistible, as they have been for the last five hundred years," writes the roboticist Rodney Brooks.[1]

"Asking whether such changes are 'wise' or 'desirable' misses the essential point," insists Gregory Stock, who titled his most recent book *Redesigning Humans: Our Inevitable Genetic Future*. "They are largely not a matter of choice; they are the unavoidable product of . . . technological advance."[2]

We are on a "journey into a rapidly evolving future that no man, no woman, could stop," writes Lee Silver.[3]

Such certainty sounds impressive. But in fact it is a bluff, a calculated attempt to demoralize the opposition. It's a high-sounding version of "You can't fight city hall," designed to stop debate before it can get going. There *are* good reasons to think that these technologies, if developed, will be used; that's what usually happens. But there are better reasons to think we may still be able to control them, that the visceral recoil from the loss of meaning that I've been describing will translate into effective political resistance. Past and present provide examples of such restraint—they provide reasons to think that democracy may yet hold a card or two.

In their effort to end the debate before it's begun, some techno-zealots argue that it's too late to draw the line: we're already engineered. People with hearing aids or artificial joints are protorobots, they insist; if you work at a computer linked to a network you've already embarked on an "unmistakably doubled articulation" that "signals the end of traditional identity."[4] Stephen Hawking, the British physicist afflicted with amyotrophic lateral sclerosis, uses a keyboard to communicate, so according to one philosopher he's not entirely human anymore: "Where does he stop? Where are his edges?"[5] If "biological manipulation is indeed a slippery slope," says Stock, then the fact that we use things like birth control means "we are already sliding down that slope now and may as well enjoy the ride."[6]

All this is more rhetorical than real, however. My mother had a knee implant last fall. She did not come out of the operating room a robot in any useful sense of the word—she's who she always was, except she can get in and out of the car a little easier. I'm typing this book on a computer, but it hasn't erased my identity. Calling someone a cyborg because he has a hearing aid is not much more useful than calling him a cyborg because he has a digging stick or a plow. Such tools fall easily within the bounds of the traditionally human;

they don't make us uncomfortable, precisely because they don't tamper with our identity at a level deep enough to matter. The people insisting otherwise are like the "pro-transhumanist activists" who last summer started sneaking into health food stores and putting stickers on the produce that read, "This product is genetically engineered by hybridization techniques."[7] They knew full well that ten thousand years of cross-breeding was different from sticking flounder genes in strawberries; they just wanted to blur that line, to intimidate their opponents.

Techno-enthusiasts stand on stronger ground when they argue that the use of these new technologies is inevitable because they will be fairly easy to develop—so easy that the kinds of controls we place on, say, nuclear weapons won't work. "The most crucial resources required to build a nuclear weapon—large reactors and enriched sources of uranium or plutonium—are tightly controlled by the government itself," writes Silver. "The resources required to practice reprogenetics—precision medical tools, small laboratory equipment, and simple chemicals—are all available for sale, without restriction, to anyone with the money to pay for them."[8] The London *Times* quoted one "senior British embryologist" who said "there are plenty of people here who would be interested. I have played around with embryos after hours."[9] Indeed, the sheer pleasure scientists take in doing something difficult will always be a motive: "It is virtually inevitable it will get used and for the most banal reasons possible—to make some money, or to satisfy the virtuoso scientists who created the technology," says Liebe Cavalieri, a molecular biologist at the State University of New York at Purchase.

Against such realities, national laws might be a flimsy bulwark. When the U.S. government set certain limits on stem cell research in 2001, for instance, several American scientists went to British universities to continue their work.[10] If Congress bans human therapeutic cloning, cautions the MIT biologist Robert Weinberg, U.S. biologists will "undertake hegiras to laboratories in Australia, Japan, Israel, and certain countries in Europe."[11] When the Raelians wanted their own cloning clinic they rented an office in the Bahamas; their chief rivals to produce the first human clone, Severino Antinori and Panayiotis Michael Zavos, reportedly set up operations in more than one foreign country. "We don't need America," Zavos said. "The world is wide open to all of us."[12] And not just the countries we think of as most advanced—the Chinese, for instance, have modern IVF clinics, and with their one-child policy perhaps an incentive to push for "better" children.[13]

If governments tried to outlaw such work, advocates warn, it would "only force the research underground, making it impossible to monitor and regulate."[14] Bill Joy's call for relinquishment of, say, nanotechnology would need to be "one hundred percent effective," says Ralph Merkle, of Zyvex. "If it's 99.99 percent effective, then you simply ensure that the 0.01 percent who pays no attention to such calls for relinquishment is the group that will develop it."[15] If we ban bioengineering, warns George Gilder in a phrase that will resonate with all NRA members, "only outlaws and rogue states will command the capability."[16] International mafiosi will set up biotech labs, warn Gregory Paul and Earl Cox, "not only for their power and money, but because crime bosses are mortals who want to escape death and will be ruthless in their search for the technologies of immortality."[17]

By most accounts, though, legal entrepreneurs and not mobsters are most likely to develop the new technologies, simply because the money is so good. Craig Venter ran up a hundred-million-dollar fortune sequencing the human genome, and the lesson was lost on no one—patent applications in cloning and stem cells jumped 300 percent between 2000 and 2001.[18] Big corporations (with big lobbying power) are placing big bets: in the words of one business journalist, "hybrid blends of high tech and biotech" are likely to become "the next industrial giants."[19] And they'll be hard to police not only because they'll contribute to political campaigns, but also because their work, day in and day out, won't be dramatic enough to attract notice. Except for a few watersheds like the first clone, the big advances will come in stages, points out Ray Kurzweil: "Every company has to innovate all the time, and so it happens incrementally."[20]

In the end, says Silver, "there is no doubt about it. Whether we like it or not, the global marketplace will reign supreme."[21]

The only alternative, they insist, is a police state; if you don't want *Brave New World,* you get *1984.* George Gilder and Richard Vigilante write that any policy of relinquishment would require a "massive regime of surveillance and regulation of businesses."[22] Patients in IVF clinics, for instance, would have their privacy imposed on to guarantee that they were getting only unenhanced embryos.[23] Indeed, some have already proposed a system of "universal genomic profiling" on every citizen—we'd be checked regularly to make sure we hadn't been enhanced, much as athletes are randomly tested for steroids.[24]

The argument is strong. Our success with prohibitions is mixed at best. Americans drank throughout the twenties, and they smoke dope today. Economic sanctions often leak. Commercial pressures often trump wise policy making. And so on. But these are arguments, not proofs—they don't guarantee that widespread use of these technologies is inevitable, merely that it is likely.

"Likely" is a long leap from "certain." In fact, there are examples from the present and the past to hearten us. We'll begin with three very different cases from around the world. Then we'll see if the same kind of restraint could be brought to bear on these new challenges.

I do not wish to drive a team of horses, or to get rid of my electric line. I do not wish to dress in black or trade in my car. Suspenders I can do without. I hesitate even to say the word "Amish," because of the images that come to mind. The rest of us do not need to be Amish *in any way*—except maybe one. The Amish are the most technologically sophisticated people on this continent, the best at picking and choosing among innovations, deciding which ones make sense and which ones don't. Their criteria aren't important for this discussion; the Amish have a particular set of notions, grounded in their religious faith, about what constitutes success. But their process, the mere fact that they are capable of discrimination, makes them the first small step in the proof that nothing is inevitable.

To call the Amish technologically sophisticated will sound odd to many ears. We're used to thinking of them as "quaint," as "out of another time," as intentionally backward. Hans Moravec says they live "in a perpetual state of early-nineteenth-century rural industrialization" that will only "erode" in the face of modernity.[25] But this is caricature and it is mostly wrong. The Amish do not reject technology; it's true that they don't own computers, but it's also true that they've allowed somatic gene therapy trials on their children afflicted with several hereditary diseases. (In 1991, a group of Amish carpenters built a special clinic for the testing and treatment of children with inherited illness.)[26] In farming, they have often adopted new machinery before their "English" neighbors—hay loaders and manure spreaders in parts of Ohio, grain binders in Pennsylvania.[27] But they don't have tractors; they have horses.

The Amish have thought long and hard about tractors and decided that tractors would erode their community. They think tractors would lead to larger and larger farms: farms that would eventually need trucks. And as horses became less useful, members would face an ever greater temptation to buy cars. Cars aren't evil—the Amish will travel in someone else's, if need be—but they help break down the fabric of social and economic life: suddenly people are living farther apart, and they have a big new expense to meet. Meanwhile, horses are cheaper, they create their own replacements, and they don't force you into dependence on oil supplies, or spare parts, or mechanics. And horses have some intangible value, too. I remember working one morning on an Amish farm in Ohio, gathering corn. Eight or nine neighbors worked with their teams in the field, and instead of the earsplitting diesel roar there was quiet enough to talk. The stalks rode a noisy mechanical conveyor belt into the silo, but the rest of the time the horses forced a relatively leisurely pace on the whole operation. The job was hard enough, but not *rattling;* when everyone gathered for lunch, they were tired but not tense.

Or take communications technology. From the very beginning, the Amish realized that telephones had real advantages. You could handle simple business without a trip to town, and in an emergency you could summon help. But they also knew that the phone lines represented a link to the larger world they distrusted—that they "disrupted the harmony of the community" by making face-to-face visits less common, and that they "disturbed the style and pace of work" by producing constant interruptions. So, over the decades, most communities worked out a clever arrangement. If you travel the back roads of Lancaster County in Pennsylvania or Holmes County in Ohio, you'll often see small shanties standing at the curb. People sometimes mistake them for outhouses, but actually they're homemade phone booths. Three or four families may share the phone, using it mostly for outgoing calls; when they need to call long-distance, they use a credit card.[28]

The point is not whether you think this is a *good* decision or not. (For you, a cell phone constantly ringing may be an affirmation of community.) The point is, it's a *decision,* made by a group of people, enforced over time by custom and habit. And a decision made for a reason. A larger community could examine, say, germline genetic engineering much as the Amish have examined television—that is, conceding that it offers certain benefits, but deciding that in the end those benefits are not worth the loss of meaning they would automatically entail. The larger society at the moment has a primitive and superstitious belief that we must accept new technologies, that they are somehow more powerful than we are. Which makes the Amish in some ways the most modern American subculture—far more modern than some fellow with a cell phone who doesn't really like it how it changes his life, but has one just because it seems "normal."

This discrimination among technologies, by the way, has not handicapped Amish society. Their numbers grow steadily and impressively—by more than 30 percent each decade, as the majority of their children decide to join the church.[29] Economically, their farms make half again as much money per acre as the average, and the Amish pay far less of what they take in for fertilizer and machinery; by incurring little debt, they weather bad crop years or low crop prices without the distress and bankruptcy so common in boom-and-bust conventional agriculture.[30] Their life is not perfect—but they prove that technological discrimination is not necessarily impossible, and does not necessarily lead to inbred stagnation. It's a power within the realm of human possibility.

History provides much larger examples of such picking and choosing. Not European history, so much—for the last five hundred years we've been in an all-out race to technologize—but the history of the rest of the world. We'll examine two cases, one from China and the other from Japan. Again they are problematic, tentative, not directly relevant to the choices we face. But they also indicate what's possible.

In 1414, a huge fleet left China. By "huge," I mean 250 times the size of the one that Columbus took toward the New World nearly a century later. Zheng He, who commanded the Chinese ships, had sixty-two galleons, any one of which could have held the *Niña*, the *Pinta*, and the *Santa Maria* on its decks. He took 30,000 people on his voyage, whereas fewer than one hundred accompanied Columbus. Zheng He's fleet visited Bengal and brought home giraffes; they called in Africa and Sri Lanka; they demonstrated a navy that was "many times larger and more powerful than the combined maritime strength of all of Europe." And yet, a few years later, the government took the conscious decision to scrap its blue-water ships and retire from overseas trade and exploration. As Samuel M. Wilson reported, "Zheng He's magnificent ships finally rotted at their moorings."[31] China renounced, in other words, the technology that would soon spread European influence around the globe—a technology as revolutionary in its time as germline engineering may be in ours.

This strange story reflects twists and turns in the battle between the two "political parties," the two ideologies of the day. On one hand, the Confucian bureaucrats, hewing to teaching two millennia old about order and virtue; on the other side, the court eunuchs, who had begun as caretakers of the imperial harem and had gradually become, "in effect, a separate echelon of administration, not unlike a present-day security system."[32]

In 1399, during a civil war between factions of the Ming dynasty, a eunuch general named Ma He defended a crucial piece of turf, the Zheng Village Dike, during what turned out to be the conflict's decisive battle. The victorious emperor, a usurper named Zhu Di, rewarded the general by renaming him Zheng He and putting him in charge of all palace construction and engineering. Soon thereafter, Zheng He was ordered to build a mighty navy.[33] Historians differ on why the emperor wanted ships: the Chinese faced continuing piracy from Japanese raiders; the Middle Kingdom wanted to reassert its position as the world's center; foreign tribute may have seemed particularly attractive after civil war had drained the national coffers; and Zhu Di may have been trying to track down the nephew he had overthrown, to make sure he never returned. But there is no doubt about the glory of Zheng He's fleet. Its seven voyages took him to present-day Vietnam and Bangladesh, to Mogadishu, Mombasa, and Madagascar, to Brunei and Chittagong, Luzon and Timor.[34] Six hundred years later, the *New York Times* reporter Nicholas Kristof, who followed the route of the voyages, found descendants of the Chinese seamen on islands off the coast of Kenya, and Indonesians who still prayed to Zheng He for cures or for good luck.[35]

But Kristof had to work almost as hard to find the abandoned grave of Zheng He on a hillside above Nanjing. The great captain had no monument and no glory. For at the height of his achievements, China turned its back on the sea. The emperor who had sent him on his voyages died in 1424, and the Confucian bureaucrats won a series of internal skirmishes; by the time they were done, not only had the navy been dismantled but laws had been passed making it a capital offense to build a boat with more than two masts.[36] Again, scholars differ on exactly why: China's Grand Canal had been rebuilt, reducing the need for coastal seafaring; the Mongols were threatening from the west and resources were needed to build the

Great Wall; the voyages had been expensive and relatively unremunerative.[37] Clearly, though, the Confucians detested the legendary corruption of the eunuchs, felt that foreign commerce was less important than building up the strength of the domestic economy, and believed that, in general, China had little to learn from the "barbarian" world outside. Kristof argues persuasively that the decision may rank as "the biggest news story of the millennium" because as a result the West, not the East, was to dominate world trade and come first to commercial and industrial might. The Chinese, he writes, "were simply not greedy enough. . . . Ancient China cared about many things—prestige, honor, culture, arts, education, ancestors, religion, filial piety—but making money came far down the list." This "folly," which China kept up through the beginning of the twentieth century, sealed the country's fate.[38]

Again, we're less interested here in whether it was a good idea than in the mere fact that it happened: that a great people turned its back on a promising technology. (It is worth noting that, having taken a pass on colonialism, the Chinese spent the next three centuries occupied otherwise than in looting and enslaving much of the globe.) And China didn't exactly stagnate as a result; it continued to pioneer fields from rocketry to ceramics to printing. More to the point, in the words of the astute China scholar John K. Fairbank, the Ming dynasty and its successors, who ruled from 1368 to 1912, mark "one of the great eras of orderly government and social stability in human history." Vast millions lived out their lives in some kind of harmony, any disruption minor "compared with the organized looting and massacres" that plagued Europe during the Thirty Years' War and thereafter.[39] The Chinese chose their definition of meaning—progress within tradition—over the pell-mell dynamism of the West.

An even more impressive example comes from Japan, a century later. The story of how that nation voluntarily gave up guns for a period of three hundred years is little known, and has been best told in a slim volume by Noel Perrin. It begins in 1543, when Europeans arrived in Japan, carrying, among other things, firearms. The Portuguese, Spanish, Dutch, and English found a highly advanced feudal society, complete with knights in armor—the famous samurai, who proved adept marksmen. Within a decade, the highly skilled Japanese craftsmen were making the new weapons in large quantities; soon they'd even improved them, adding a box-shaped matchlock protector to enable firing in the rain.[40] The Japanese were famous for their swords; the best blades were hammered and folded and rehammered till they could cut through tempered steel armor.[41] But guns could kill at a distance, swords could not, and by the late sixteenth century, "guns were almost certainly more common in Japan than in any other country in the world."[42]

But then, with the rise of the Tokugawa shogunate in the early 1600s, Japan did away with firearms. Their manufacture gradually ceased; the samurai stopped using them in battle; they disappeared so thoroughly that when the Europeans returned in the nineteenth century they found, instead of shore batteries, a huge sheet of cloth painted to look like a fort bristling with cannons.[43] Perrin provides a number of practical reasons for the decision: for instance, the Japanese islands were remote and by their nature hard to conquer, so the residents had little fear of invasion. And indeed, guns were just one of the foreign ideas the Japanese rejected: they ran off the Christian missionaries, too. But the deeper reasons had to do with meaning. To the Japanese samurai, who made up perhaps 10 percent of the population, "the sword was not merely a fighting weapon, but the visible form of one's honor." Unless a man had been granted the right to wear a sword, he couldn't even have a family name. Even when gunmaking was at its height in Japan, in the late 1500s, the government honored its four leading gunsmiths by giving them . . . swords. An elaborate aesthetic surrounded the use of swords, governing the elegant and graceful way one should move one's body.

In the end, the samurai simply felt that guns were crude, that any peasant could use one, that they were destroying the intricate architecture of honor and civility that had marked even the nastiest samurai warfare. A battle between sword-wielding armies actually consisted of a large number of single conflicts—warriors introduced themselves, and then set about fighting. "Such a battle could produce almost as many heroic stories as there were participants."[44] Gunfire, by contrast, offered only anonymous slaughter. And so, for three centuries, until the arrival of Commodore Perry left them no choice, the Japanese dispensed with perhaps the world's single most decisive technology. (It didn't, obviously, take them long to catch up.)

Though they resisted the innovation of firearms, it would be wrong to imagine that Japan stagnated under Tokugawa rule. With a population larger than any European country's, the Japanese supported five universities, the smallest of which was larger than Oxford or Cambridge.[45] Even as they were phasing out guns, they built the world's first great waterworks, to supply Tokyo; agriculture and mining flourished; public health measures freed the country from scourges like typhoid. And this at a time when plague and constant warfare had cut in half the population of some European countries.[46] In the words of Edwin O. Reischauer, the great Japan scholar of the postwar era, "this was probably the longest period of complete peace and political stability that any sizable body of people has ever enjoyed. And yet it was a time not of stagnation but of very dynamic cultural and economic growth."[47]

I am well aware that these cases don't prove that we will be able to hold human genetic engineering or aggressive nanotechnology in check. They come from other ages and other cultures; they were confined to one society; they occurred under authoritarian rule; and they eventually gave way before the force of competition from abroad. They do, however, demonstrate one thing. In Perrin's words, we sometimes think of "progress" as "something semi-divine, an inexorable force outside human control. And, of course, it isn't. It is something we can guide and direct and even stop."[48]

We may not be Tokugawa Japan or even China, and most of us aren't Amish. But in certain ways, we have advantages over them when it comes to reining in these new technologies. Most important, *the technologies don't yet exist.* The genie's poking her head from the bottle, but her shoulders are too wide for her to easily slip out. She's singing a siren song, but there's time yet (if barely) to stop up our ears.

Not only that, but we have a surprisingly good track record in recent decades at just this sort of control. Take nuclear weapons, for instance, one of the first technologies that threatened to erase meaning (and everything else). Since Hiroshima and Nagasaki, scientists, diplomats, and many ordinary folk have fought to rein in nuclear weapons. They've succeeded in certain respects (the superpowers have begun to shrink their arsenals) and failed in others (the weapons have spread to new nations), but so far the bomb's been dropped just twice. Biological and chemical weapons have existed in arsenals around the world for decades—some of them are relatively easy to make ("the poor man's nuke")—yet efforts to control them have been mostly successful. Saddam Hussein gassed dissidents; someone mailed anthrax to Americans; but our systems of surveillance have by and large done the job.

These analogies are imperfect: almost no one *wants* nuclear war, for example, while some people surely will want to program their children. But we've also turned our backs on more popular technologies. DDT, for instance: when Rachel Carson inconveniently pointed out that it was ravaging wildlife, people eventually overcame the big money behind it and enacted an effective ban. The same chemical companies that fought Carson also fought efforts to protect the ozone layer; if we banned chlorofluorocarbons, they suggested, black marketers in foreign countries would just produce it. Indeed, there's been a little of that—auto shops buying CFCs for air-conditioning from clandestine factories in the former Soviet Union. But it hasn't been enough to matter: the atmospheric concentrations of the chemicals are declining, and the ozone layer has begun the long task of repairing itself.

Or consider genetically modified crops. It's true that GMO corn and soy spread quickly across the American grain belt with hardly anyone noticing; corporate power made sure that Washington wouldn't regulate the new varieties. But across Europe, consumers began to say in large numbers that they simply wouldn't eat the stuff. Grocery distributors stopped buying it. The growth of GMO foods was slowed, and in some cases reversed. Worldwide outcry against the so-called Terminator gene, from which the plants that grew were, each and every one, sterile, and which would thus have ended ten thousand years of seed-saving, forced Monsanto to drop plans for its commercialization.

Even in the world of sports, that out-of-control drug playground where we began this discussion, real progress has been made. By most accounts, the Tour de France was markedly cleaner in 2001 and 2002 than in preceding years; people are winning discus events with shorter throws, a certain sign that steroid intake has dropped. Similar results are appearing in other competitions as well—medical tests show that the blood of the average elite cross-country skier is thinning as skiers give up on drugs like EPO.[49] One sports federation after another has instituted lifetime bans for anyone caught cheating, and in some cases that's been enough to scare athletes and coaches straight. Even if the genie manages to squirm out of the bottle and you can't force her back in, sometimes you can build a pretty tight cage around genie and bottle both.

And the new human genetic engineering has features that make it even more susceptible to a relatively effective ban.

For one thing, this engineering is relatively difficult. Proponents talk all the time about how it's simple lab work, and just one step up from in vitro fertilization, and so on. In certain ways they're right—but genetic engineering is not like mixing methamphetamine in a kettle on the stove. Why does Rael not have a clone yet, despite his small army of willing surrogate mothers? His cloning director, Brigitte Boisselier, set up her first lab in an empty room she rented for $350 a month in a high school in the rural West Virginia town of Nitro. She told the National Academy of Sciences and Congress that she was on the verge of creating a clone, but when federal investigators took a look at her lab they found

> a blackboard and a desk. On the desk sits a computer hard drive, keyboard, scanner, printer, and cordless telephone. There are two chemistry lab tables topped with a water filter, scales and other machines. They are equipped with sinks, a roll of paper towels, a box of rubber gloves, a near-empty bottle of Softsoap and a half-full bottle of Palmolive. There is a filing cabinet, a refrigerator, and an incubator. And posters with blown-up pictures of eggs, human or animal, adorn the walls.

Boisselier was, they decided, "many years away from attempting to clone a human."[50]

And what about her competitors Zavos and Antinori, who promised in the fall of 2001 that they would have "cloned embryos in the very near future. That is, three or four months from now"?[51] Just as the advocates predicted, the two shopped around for countries that had yet to ban reproductive cloning. Speaking outside a congressional hearing, Dr. Zavos claimed to have set up two overseas labs, one in "I guess you could say it's Europe," and the other in "territory between Greece and India." Still, he acknowledged, it has been a little hard "tiptoeing through the tulips where all these countries and governments are throwing spitballs at you."[52] For a while, speculation centered on Israel, and the German newsmagazine *Der Spiegel* reported that the lab would probably be in the coastal resort of Caesarea (which would give new meaning to the term "cesarean"), but a few weeks later Antinori announced at a medical conference in the United Arab Emirates that in fact he had succeeded in impregnating a woman with the cloned child of "a wealthy Arab." "Imagine," he told a colleague, "it has been possible to carry out in a Muslim country a kind of research that was impossible to do in the West."[53] A few weeks after that, Dr. Zavos said he was splitting with Dr. Antinori, and that the Italian not only didn't have a clone, he also had no lab, no patients, and no doctors to work with him. But, said Dr. Zavos, *he* would definitely produce a clone sometime soon, because he was working with other researchers that he identified only as "the Michael Jordans of the business." He refused to say exactly when the baby might be born. "We are not in the business of delivering a pizza," he said.[54] By November 2002, Dr. Antinori was back making the rounds of the talk shows, this time with an unidentified woman who he said would bear a clone in January of 2003.[55] Meanwhile, Dr. Boisselier and the Raelians said they had five pregnancies under way.[56]

Eventually, perhaps by the time this book rolls off the printing press, they—or somebody—*will* succeed. They will clone someone, or they will perform some genetic enhancement. As we've seen, all this and more has already happened with animals. But even if a few scattered labs manage the trick, that won't necessarily change the world—the engineered human is not like a nuclear bomb, where one is a lot. Rael can open the debate, but not close it. "In all likelihood, some people will become genetically enhanced," the law professor Maxwell Mehlman wrote recently. But "if the number of enhanced individuals is sufficiently small, we may be able to ignore them."[57] They would be, by definition, "freaks." Something large enough to matter in the real world would require big money: big money, chasing big payoffs. Even such a stalwart of the bioengineered future as Gregory Stock wrote in 2001 that "human germline manipulation could never hope to become technically feasible except as a spinoff of huge research projects in related arenas. Any financial payoff from it is too distant, liability issues are too worrisome, and development costs are too high."[58] In other words, it will take government money and corporate money to make such schemes work on any significant scale, and that's precisely what would make it possible to slow such work down, to direct and guide it. If it's illegal, your local IVF clinic won't be doing it, simply for fear of enormous lawsuits when things go wrong. Maybe the superrich will be able to fly to a clinic in Brunei; but maybe no country would be able to afford the sanction of the rest of the world, were that sanction sufficiently strong. We *should* fear Gentopia, Inc., or whatever the company with the patents will be called; but it should fear the political world, too, for without its tolerance there will be no access to capital or markets.

The idea of inevitability is a ruse, an attempt to preempt democratic debate. There will be a contest. And it begins with real advantages on both sides. The would-be genetic engineers wear the cloak of "progress," and if it's a little more tattered now than in the past, it's still pretty impressive. They represent a technology that could make large amounts of money. In the laissez-faire economic world we now inhabit, they can go right ahead if no one says, "Stop." And they are, by definition, extremely smart.

On the other hand, they must contend with a gut revulsion at the prospect of "designer children," an instinctual allergy to this new level of interference. Poll after poll shows that people don't like cloning, and that to the extent that they understand human germline engineering they like that even less. All the European nations have already explicitly banned germline manipulation; the United States, as Rich Hayes puts it, is the "rogue nation" on these questions. This is a fight that should be decided like other fights—over nuclear power, or privatization of Social Security, or any of the thousand other issues that democracies must debate and decide. It's an *issue,* not a fact.

The fight is already well under way, around the world and especially in the United States. Any doubts on that score disappeared in the spring of 2002, when, in the middle of the congressional debate over cloning, Harry and Louise suddenly reappeared. This fictional middle-aged couple first sprang to life in 1993, when the health insurance industry used them in a series of ads to torpedo the Clinton health plan. Now the same ad agency was using the same pair sitting around the same kitchen table to argue that there should be no restrictions at all on "therapeutic cloning."[59]

The ad wasn't the first attempt of the genetic engineers to "buy mindshare." As early as 1997, Burson-Marsteller, the world's largest public relations firm, was advising a consortium of pharmaceutical companies that they needed to spread "symbols eliciting hope, satisfaction, caring, and self-esteem."

They must have been pleased when, a few years later, the American Museum of Natural History, in New York City, hosted a huge exhibit called "The Genomic Revolution." It had been suggested by a museum board member, Dr. Frederick Seitz, and was funded by his Lounsbery Foundation. When the journalist Jackie Stevens asked Seitz why he'd done it, he said, "Enthusiasm for [genetic technologies] needed to be boosted a bit." And doubtless it was. Visitors entered the exhibit through "a dark room aglow with video loops of talking heads refracted through Plexiglas, seemingly coming from nowhere." All around, signs insisted that, for instance, "by the year 2020 it is highly possible that the average human life span will be increased by fifty percent." When Stevens interviewed Rob De Sale, a molecular biologist and the museum's curator, he told her he didn't necessarily believe the claims, but that "veracity wasn't his intention. 'It was designed to get people to turn the corner,'" he said.[60]

Perception management is always key when you're selling something people are wary of. Just as genetically modified foods gave way to "biotechnology," so various advocates have suggested that "therapeutic cloning," which is already a euphemism of sorts, might better be called "nuclear cell transplantation to produce stem cells."[61]

More fundamentally, however, the techno-enthusiasts keep nudging the goal posts. Fifteen years ago, when all human genetic engineering was suspect, researchers pushed for the right to do somatic gene therapy. As we've seen, this is mostly a sensible idea, and one of the ways that scientists got permission to do it was by drawing a sharp line between it and germline engineering. But it turns out that at least some of them were behaving tactically. As James Watson, who has participated in the entire process, recalled in 1998, "Partly it was in order to get somatic therapy going that it was said, 'Well, we're not doing germline. That is bad. But somatic is not bad morally.' That virtually implied there was a moral decision to make about germline, as if it was some great Rubicon and involved going against natural law."[62] But no sooner had they won permission for somatic work than they were back asking for the right to tinker with the germline, saying the Rubicon wasn't a Rubicon after all, just one more little stream. The 1998 conference "Engineering the Human Germline," at UCLA, was designed, according to its organizer Gregory Stock, to make germline engineering "acceptable" to the public.[63] No critics were invited to speak; instead, a large crowd that included journalists from most of the nation's major media soaked up copious doses of cheerleading. Any international attempt to regulate germline engineering would be "a complete disaster," said Watson.[64] "The question is not if, but when and how," the conference report stated.

All in all, the campaign has met with considerable success. By January 2000, *Time* had devoted a breathless cover story to medical biotechnology; Lee Silver contributed a long commentary explaining the inevitable wonders to come. An edition of *Nightline* was also typical: the generally skeptical Ted Koppel interviewed Silver about germline engineering, and then said, "If we spent our lives worrying about the impact of unforeseen consequences, we'd never get anything done. We'd be paralyzed." As the show ended, Koppel tossed off a cheery homily: We don't need to worry too much about this new technology because "Mother Nature" will "cause us to adapt to change" so that any genetically engineered elite will be challenged by "some of those other kids who are going to adapt by staying home and studying harder."[65] Well, now that that's taken care of . . .

Journalists are, like the rest of us, intimidated by scientists. In fact, the basic argument against restrictions of any kind often goes like this: We're scientists. We understand this better than you. Leave us alone.

Consider, once more, Michael West, a man who is obsessed with mortality, the one who cloned the first human embryos. When he's talking with the *New York Times,* he "insists on an open national debate about stem cell research."[66] But when he relaxes in friendlier forums—for instance, in an on-line interview on the topic of "living forever"—he's more forthcoming. "In regard to your question about trusting policy to politicians," he says, "I am increasingly of the opinion that we need a specialized advisory group *with some leverage* [italics mine] to advise the Congress on these issues. It's disgusting to me when members of Congress propose legislation" to ban such work "and the author of the bill can't even demonstrate that he knows what he is trying to criminalize." If there are any doubts about what West means, he goes on to describe his fellow researchers as "modern-day Galileos" and to say that "given the rate of new developments in biotechnology, and the need to rapidly assess trends, I suspect we really should have a national science authority. Although I recognize the weaknesses of bureaucracy, it would be better than having insurance salesmen from who-knows-where pontificating on such important issues."[67]

Since we live in a system that allows insurance salesmen from who-knows-where to stand for election and then decide the issues we face, that's saying rather a lot. But West is not alone. "What kind of country do we live in when we are talking about banning research?" asks Gregory Stock. "Science should not be micromanaged by politicians."[68] An editorialist in *Nature* put it this way: "Questions that seem cut and dried to professional people may be deeply worrying for the more general public. . . . It would be a shame if biology's megaproject, with all its promise, encountered the incoherent opposition that has blocked the path of great endeavors in other fields."[69] Marvin Minsky, the father of modern robotics, writes that democracy "works well for easy problems," but not "when the issues get too complicated for laymen to understand." Indeed, even when Minsky tries to explain artificial intelligence to regular folk, he finds himself frustrated. "I find that in order to have any effect at all, I have to spend fully half of my allotted time explaining the simplest facts of basic science. In other words, just to establish the most basic kind of communication, I find it best to say a little about what Newton and Darwin discovered, or Franklin, or Freud, or Turing."[70]

We are all tempted to leave things in the hands of wiser people who manage to make troubling phenomena—designer children, robots merging with people—seem benevolent, inevitable. It's always nice not to have to think. But it would be extremely unwise to let scientists walk away with that much power. In the first place, non-scientific concerns can cloud scientific judgment. Money, for instance: biotech has become one of the strongest magnets for venture capital in recent years, and any biologist worth his centrifuge is awash with stock options. In 2001, "a survey of medical experts who write guidelines for treating conditions like heart disease, depression, and diabetes has found that nearly nine out of ten have financial ties to the pharmaceutical industry, and the ties are almost never disclosed."[71] The *Lancet,* Britain's leading medical journal, wrote about how groups like the American Heart Association were accepting eight-figure "donations" from companies like Genentech even as they were writing treatment guidelines recommending products from the company.[72] Remember that young volunteer, Jesse Gelsinger, who died in a University of Pennsylvania gene therapy experiment? After months of digging, a pair of journalists learned that the director of the study had a financial interest in its outcome—and that he hadn't shut it down even when four previous volunteers had suffered serious side effects.[73]

But there's a deeper reason to give no special weight to the judgment of scientists: in truth, they have no better idea how we should proceed than anyone else who's thought about these issues. The devil is not in the details; it's the basic *thrust* of these technologies that's diabolical. Understanding which chromosomes are responsible for the expression of which proteins doesn't give you any added insight into whether designer babies are a good idea, any more than figuring out how to make an atom bomb turns you into an expert on when or where you should drop it. Those decisions call for completely different sets of skills, something scientists usually realize. The thousands of climatologists studying global warming, for instance, have produced several five-year updates for policy makers, outlining exactly what we know and evaluating the effects of different approaches to controlling carbon dioxide. But they've been scrupulous in letting governments decide which taxes to raise or which technologies to ban. Such decisions they leave to dictators and kings and, in countries lucky enough to be democratic, to all of us.

In the case of these new technologies, however, the prizes are so tantalizingly large, the science is so interesting, that researchers have claimed more authority than they deserve. If you think I exaggerate about the lack of special wisdom conferred by mere scientific genius, I have two words for you: James Watson. Every single time there is an announcement or a press conference or a commission on anything genetic, James Watson is up on the dais. He holds thirty-two honorary degrees. And why not? In 1953, he and Francis Crick discovered the double helix. They *figured it out.* His brain is exceptionally sharp in that particular way that allows for scientific discovery.

But does that same brain produce any special insight into whether it makes sense to use those discoveries for the manipulation of children in the womb? Consider just a small sampling of his offerings: When asked if he feared that genetic engineering could be used for eugenic ends, he replied, "It's not much fun being around dumb people."[74] At a Berkeley lecture in 1999, according to an account in the San Francisco papers, he startled the audience by explaining that thin people are unhappy and therefore more ambitious. "When you interview fat people, you feel bad, because you know you're not going to hire them," he went on. He also showed slides of women in bikinis and veiled Muslim women to "suggest that controlling exposure to sun may suppress sexual desire and vice versa."[75] (Much of his recent autobiography concerns the "girls" in his life. He describes his usually fumbling assignations in some detail; reviewing the book for the *New York Times Book Review,* Barbara Ehrenreich noted that "the man who did so much to elucidate reproduction at the molecular level seems to have had little idea of how to go about it himself.")[76] Speaking in Toronto in October 2002, he said germline engineering could be used to benefit the shy, the hot-headed, and "cold fish."[77] This is the man who "strongly favors controlling our children's genetic destinies," who believes "one should never put off doing something useful for fear of evil that may never arrive."[78] Some people, he opined, "are going to have to have some guts and try germline therapy without completely knowing that it's going to work," if for no other reason than to "cure what I feel is a very serious disease—that is, stupidity."[79] "Guts" are a favorite topic: "No one really has the guts to say it. . . . I mean, if we could make better human beings by knowing how to add genes, why shouldn't we do it?"[80] "Better"—what does that mean? How should we judge it? What matters about a human being? "I'd like to give up saying *rights* or *sanctity*," Watson explained. "Instead say that humans have needs, and we should try, as a social species, to respond to human needs—like food or education or health—and that's the way we should work. To try and give it more meaning than it deserves in some quasi-mystical way is for Steven Spielberg or somebody like that. It's just plain aura, up in the sky—I mean, it's crap."[81] The man's entitled to his opinions. He should get his vote just like everyone else. But I'll take my chances with the insurance salesman from who-knows-where.

If we don't dare abdicate the power to make decisions to research scientists, then perhaps we could turn to doctors. We're used to deferring to their judgment: if they produce a patient and say, "This person might be helped by this procedure," it's tempting just to go along. But as we've seen, there are many other routes toward treatment of genetic disorders that don't entail species-changing revolutions. And, in fact, it's the doctor's admirably total dedication to her *individual* patients that makes her just the wrong person to figure out what makes sense for the society. The dedication to an individual leaves little room for larger considerations; as anyone who's ever watched a parent kept "alive" in a tangle of machinery can attest, sometimes doctors can't even think clearly about the deeper needs of their own patients.

Doctors' struggles with deciding when to stop intervening against death helped lead to the creation of a makeshift profession, "bioethics" Though "anyone who cares to hang out a shingle"—a lawyer, a philosopher, an anthropologist, a doctor, a theologian—can join, "bioethicists" are nonetheless "quoted almost daily in the media, testify before Congress, and advise the president." They are, in the words of the journalist Nell Boyce, a kind of "secular priest"—and they are the last line of defense for those who would like "someone else" to make these decisions about these new technologies.[82] But that, alas, is an idle dream as well.

For one thing, all sides in these debates have "their" bioethicists, whom they use for cover. When President George W. Bush appointed a panel to advise him on cloning, for instance, he was widely accused of stacking it with conservatives who would tell him what he wanted to hear. In general, however, bioethicists have been captives of science and industry, for a variety of structural reasons. One, perhaps the least important, is simply self-interest: all the big biotech companies have ethics boards, and some of them pay their advisers with stock options or hand them checks for $2,000 a day.[83]

In a broader sense, bioethicists come from the same intellectual background as most of the people whose projects they're supposedly judging. Daniel Callahan, the president of the Hastings Center on bioethics, remarks that "most of those who have come into the field have accepted scientific ideology as much as most scientists, and they have no less been the cultural children of their times, prone to look to medical progress and its expansion of choice as a perfect complement to a set of moral values that puts autonomy at the very top of moral hierarchy." Most bioethicists thus don't ask questions about society or the species—they ask questions about individuals. Is this drug trial safe? Did everyone sign the informed-consent form? "Bioethcists have, on the whole, become good team players," says Callahan, "useful to help out with moral puzzles now and then and trustworthy not to probe basic premises too deeply."[84] So, for instance, when Advanced Cell Technology was planning to clone human embryos, the chairman of its ethics board outlined the "extreme precautions" the company was taking: the eggs would be kept in a secure location, access to which required permission from two ACT technicians; the eggs would be repeatedly counted, photographed, and videotaped; and so on. Nothing untoward would happen to the eggs—but the society that was about to be dramatically changed by this new development would have to fend for itself.[85]

Even the language of moral inquiry quickly turns technical. The sociologist Barbara Katz Rothman describes how, early in the history of genetic engineering, one group of theologians began by worrying about people "playing God"—that is, taking control over parts of life that they felt people had no business controlling. When they expressed their concerns to President Carter, he formed a commission of ethicists—but in draft after draft of their report they kept reducing the moral questions to technical ones. "Playing God" was too vague; it was translated as "acting without knowing the consequences, taking risks." So the response was to control risks—to put more filters in the lab's safety hoods; to write better informed-consent forms. The moral dilemma became somehow manageable.[86] The question isn't whether "playing God" is a sufficient objection; I've tried to show that bioengineering raises fundamental issues of meaning that don't depend on one's theology. But "ethicists" rarely do more than scratch the surface.

And how could they do any more? There are no all-knowing, all-seeing oracles to tell us the truth about these technologies. I think genetically engineering our children will be the worst choice human beings ever make—but I've explained *why* I think that, just as I would if I were writing about welfare reform or nuclear power or drug addiction. And I understand, intellectually if not emotionally, why others might want to engineer "improved" children; they weigh the costs and benefits differently. When we're deciding about a tax increase, we don't consult "tax ethicists." We listen to economists who offer predictions about effects; we look at our schools and roads and judge their condition; we look at our paychecks and gauge what we can afford; and then we go vote. All that thinking may eventually boil itself down to an ideology—we become generally "anti-tax" or "pro-spending"—and that's all right, just as it's all right to feel generally queasy about genetic manipulation or generally gung-ho about technology. But you can't take a pass on the work of decision making, can't hope that some wizard from Oz will appear to explain in a booming voice precisely what should be done. "Forums and commissions can't settle these disputes," writes the essayist William Saletan. "Nor can ethicists. It's not their job. It's yours."[87]

Happily for us, we have a system for dealing with competing ideas. It's called politics. As always, we're going to have to make choices, basing them on some mix of knowledge and thought and intuition. We will have to *choose.*

True hard-core libertarians are few in number; the party's candidate for president doesn't win many votes, in part because people deeply devoted to individualism tend to be bad at organizing in groups. But if you held the election among high-tech CEOs, he'd have a fighting chance.

Paulina Borsook, a technology journalist, published a remarkable book titled *Cyberselfish* a few years ago, right at the height of the Internet boom. She began to chronicle Silicon Valley libertarianism when she noticed "maybe for the tenth time" a particular kind of personal ad in the local paper. "It didn't say he was buffed or liked walks along the beach or was into caring and sharing. . . . Instead, 'Ayn Rand enthusiast is seeking libertarian-oriented female for great conversation and romance. I am a very bright and attractive high-tech entrepreneur.'" At technical conferences and trade shows and in Santa Cruz barrooms, Borsook spent the next months "trying to make sense of the libertarianism I found all around." Some of it was pure Randian capitalist devotion; some of it an almost New Age belief that these new technologies were so complicated that they defied regulation ("The Economy Is a Rainforest" read one Bay Area bumper sticker). And some of it came from the simple overweening pride of the techies in their own amazing dynamism—the hard work, the serial bankruptcies, finally the stunning success. (It also had something to do with how most of these men spent their working days: "as the sole commander of one's own computer.")[88] A lot of it was also pretty dumb, of course—the Internet, after all, began as a government-funded invention.

But if you read the bulletin boards where true believers in the emerging technologies congregate, it's clear how influential this school of thought has become. There's a profound conviction that trying to direct or regulate or slow down any of these species-changing projects is despotism of the worst sort. "Any attempt to stop the Extraordinary Future, even if democratically decided, will be a form of tyranny, trapping minds in human bodies when alternative venues are available," writes one futurist.[89] "Just as Winston Churchill identified an Iron Curtain of totalitarianism that was falling across Europe half a century ago, I see an Iron Triangle of opposition to meaningful progress in the human condition vying for control of the cultural scene," another speaker told a 2001 gathering of technophiles.[90] When I published an op-ed piece in the *Times* questioning cloning, my e-mail in box filled within an hour. "The imposition of moral strictures on one's fellows to me is a *vile repudiation of everything American,*" ran one of the more reasonable notes.

When this libertarian streak is applied in normal, everyday politics, most of the overstatement drops away, being replaced by the constant, mantralike use of one of the most seductive words in the modern vocabulary: "choice."

In the endless buffet line that constitutes modern consumer culture, we've learned to think of choice as our highest value. We have five hundred channels—we can choose what we want to watch. We have twenty thousand new products annually in the supermarket—we choose what we want to buy. We have every compact disc on earth trading constantly across our hard drives—we choose what we want to hear. We can get anywhere we want in eighteen hours on a plane—we choose where we'll live. We have access to every culture on the planet—we choose how we want to dress, whose food we want to eat, whose tribal jewelry we want to copy. Why shouldn't we choose what we want our kids to be like? If I want my daughter to stand six feet tall, run two-hour marathons, and remember every e-mail address on Yahoo!, how is that any different from deciding I want to dine on Honey Nut Cheerios? Gregory Stock has proposed changing the name of germline engineering to "germinal choice technology."[91]

The advocates of these technologies are at pains to announce that they don't believe in "eugenics," in anything that smacks of Nazis and Aryans and government policies. Instead of Huxley's vision in *Brave New World* of a "worldwide political state" controlling our breeding, Lee Silver stresses that "it is individuals and couples—Barbara and Dan and Cheryl and Madeleine and Melissa and Curtis and Jennifer, *not governments*—who will seize control of these new technologies."[92] And he's right. Strain as hard as you will to hear, there's no sound of jackboots clomping up the stairs. Most of our governments have spent the last few decades falling over themselves to disappear, to get out of the way of corporate and individual choice. In fact, say the proponents, the politicians will be behaving in totalitarian fashion only if they try to *stop* anyone from breeding his own little Einstein. As Dr. Watson remarked not long ago, "I don't believe we can let the government start dictating the decisions people make about what sorts of families they'll have."[93]

These positions are attractive to the left as well as the right, for liberals have their own reasons to fear overreaching by the government. The "pro-choice" movement, after all, is what we call the campaign for abortion rights; one of its most popular slogans has always been "Keep Your Laws off My Body." It was the California Democrat Dianne Feinstein who inserted a letter from biotech giant Genentech into the *Congressional Record* arguing that cloning regulators shouldn't trifle with "the legal rights of persons to free expression and inquiry in the private market."[94] (Feinstein later joined with Ted Kennedy to lead the fight against any restrictions on "therapeutic" cloning.) Gay people rightly detest the long history of government efforts to limit their choice of whom to love.

All in all, in other words, choice is a powerful rhetorical device. It is for biotech, and it will be for the other species-changing technologies that are following behind it. ("I think it boils down to the question of choices," says the nanotech pioneer Ralph Merkle. "If you have better technology, then you're no longer constrained in the range of choices of what you can do.")[95]

Which is why it's so important to say: *These are the most anti-choice technologies anyone's ever thought of.* In widespread use, they will first rob parents of their liberty, and then strip freedom from every generation that follows. In the end, they will destroy forever the very possibility of meaningful choice.

To understand why, imagine what will happen when the first few hundred parents on New York's Upper East Side decide that they will indeed spend some of their spare cash on upgrading their offspring. Almost immediately, at precisely the moment the first cover story on the subject appears in *New York* magazine, every other well-off couple of childbearing age in Manhattan will be forced to decide whether, like it or not, they're going to have to follow suit. If not, their kids may *lose*—may not get into the right preschool, may not get into Brearley, may not get into Harvard. What choice will those parents have? Only the choice to keep up with the neighbors, or the choice to put their kids behind from the start. A "choice" that will spread from Manhattan to Scarsdale, and so on down the line, as the enhancements get cheaper and as the competition gets more heated. Remember the words of Lester Thurow, whom I quoted in the first chapter of this book: "Suppose parents could add 30 points to their child's IQ. Wouldn't you want to do it? And if you don't, your child will be the stupidest in the neighborhood."[96] That's your choice.

We can see such "choices" in action already. Consider sports. Very few kids grow up thinking, "I'm going to do steroids so I can hit home runs." But at some point young athletes reach a level at which a couple of other kids are sticking needles in their butts to build their biceps, and as a result they're hammering it over the fence, and as a result they're moving up to the majors. You have a *choice,* sure. But really it was only the first few guys who had a *free* choice. In the words of one Olympic coach, "Most of the athletes didn't really want to do drugs. But they would come to me and say, 'Unless you stop the drug abuse in sports, I have to do drugs. I'm not going to spend the next two years training—away from my family, missing my college education—to be an Olympian and then be cheated out of a medal by some guy from Europe or Asia who is on drugs.'"[97]

Eventually, even the possibility of choosing may all but disappear. Say you have a family tendency toward fatness. Many other people have started to engineer it out of their children, to the point where the insurance company decides to make such a "choice" a prerequisite for coverage. Eugenics in that kind of world won't take a Hitler; as the bioethicist Arthur Kaplan puts it, it will only require people saying, "You can have a kid like that if you want, but I'm not paying."[98] Even if legislation could prevent such a scenario, what happens when your unenhanced child grows up and goes to get a job? "On the merits," maybe he gets to take out the trash. Hey, but it's your choice. No one's making you do it.

Bad as it is, however, this kind of coercion is barely half the story. The person left without any choice *at all* is the one you've engineered. You've decided, for once and for all, certain things about him: he'll have genes expressing proteins that send extra dopamine to his brain to alter his mood; he'll have genes expressing proteins to boost his memory, to shape his stature. He'll be putty in your (doctor's) hands. Since embryos, even enhanced ones, can't sign informed-consent forms, you'll be taking this on your own shoulders, exercising infinitely more power over your child than your parents did over you. Sure, they tried to raise you a certain way. They sent you to a particular school, tried to pick your friends, shared with you their prejudices. But you could walk away from some of that; doubtless you did. Maybe you turned your back entirely. *Of course* the effects of your upbringing linger—that's what it means to be a social species. But that's nothing like choosing your kid out of a catalogue; the engineered child won't have that same ability to walk away from you. If you get the proteins right, it may never occur to him to do so. He may have no more choice about how to live his life than a

Hindu born untouchable. His life may be *better* than an untouchable's, and it may be better, by some measurements, than if he hadn't been altered. But he'll have no choice.

Occasionally some of the genetic engineers will recognize this paradox, and propose that perhaps they should install artificial chromosomes that the recipients could decide to switch on later, or switch off, or modulate. But, as Leroy Hood, one of the pioneering researchers, told the UCLA conference, "it's quite clear that if we get into engineering more complicated traits, it's not going to be possible to simply make them all reversible." For the most part, the enthusiasts just throw up their hands and repeat the same old saws: "I didn't really have an option about whether I should go to school or not," says Daniel Koshland.[99] But, in fact, past a certain point, he did. His temper, his abilities, his discipline, his focus—they were all *his,* in the way that we've reckoned "his" for all of human history. They came from some combination of heredity, environment, and chance. They didn't come from a checklist. "Freedom cannot be defined solely as individual, parental choice," Carl Pope, the head of the Sierra Club, said in a speech in 2001 to the National Abortion and Reproductive Rights Action League.[100] People shouldn't be allowed to choose things this deep for their children (and for every generation thereafter).

That will involve limiting freedom, just as forbidding people to drive their cars the wrong way down a one-way street limits freedom. The liberty of one generation, ours, would be in some small way constrained (though no more constrained than that of any other generation, which never had this choice) in order to protect the far more basic liberties of those yet to come. To demand this right is to make a mockery of liberty. It's to choose, forever, against choice.

If "choice" appeals to people who see the world as a collection of individuals, a different set of arguments for these new technologies attracts those on the left, those who tend to see our society in terms of groups. For many of them, a lot of the sixties suspicion of technology has faded, replaced by a fascination with new tools. Old icons like Buckminster Fuller (who once explained that "man is a self-balancing 28-jointed adapter-based biped, an electro-chemical-reduction plant, integral with the segregated stowages of special energy extracts in storage batteries") have emerged as at least as prophetic as voices like Rachel Carson.[101] Left-wing technotopians spend little energy thinking about *more;* they'd prefer to engineer us all to be *better.*

Some of this seems fairly innocent. The philosopher Peter Singer wrote a small book in 2000 with the title *A Darwinian Left: Politics, Evolution, and Cooperation,* which argued that we have biologically rooted tendencies toward selfishness which make it hard to build a just society. Therefore, he recommended some ill-defined program of deliberately cultivating altruism.[102] Peter Sloterdijk, a leading German philosopher, sounded a more sinister note in the late 1990s when he broke a national taboo by using the Nazi-era term *Selektion* to make his case that perhaps the time had come for "biotechnological optimization." Education, he said, no longer held out much hope for civilization; it was overwhelmed by electronic media and a rising "barbarism" exemplified by American school violence. "What will domesticate man when humanism fails as a school of domestication?" he asked, leaving in the air the distinct possibility that the answer was genetic engineering in the "coming era of species-political decisions."[103]

The cleverest argument of this kind, because it took a less obvious twist, came from an American academic, Donna Haraway, who was writing in the early 1980s in the little-read pages of magazines like the *Socialist Review.* In its earliest drafts, her article "The Ironic Dream of a Common Language for Women in an Integrated Circuit: Science, Technology, and Feminism in the 1980s" is mostly pretty dull stuff, full of demands that the Democratic Socialists of America "develop a coherent science and technology policy" and that socialist feminists should work with "the Congressional Black Caucus, and California Congressperson Ronald Dellums in particular . . . to formulate alternative budgets."[104] But toward the end of the paper, she declares approvingly that people are becoming part machine—becoming "cyborgs." Haraway rewrote the essay several times, publishing it under the title "A Manifesto for Cyborgs," and before long there was a hot new field of academia. "Cyborgology," claimed the editor of *The Cyborg Handbook*, "has become a central concept for many academics, not only people in science and technology studies, but also political theorists, military historians, literary critics, human-factors engineers, computer scientists, medical sociologists, psychologists, and cultural observers of all types."[105]

Though the term was old—dating back at least to a NASA study from the early 1960s about how to "engineer man for space"—Haraway's take on the subject was novel. She wanted to discard a notion that underlies one strain of feminism: that women are "organic," "natural," and so forth. In her view, such ideas merely make repression by "rational" males easier. But if everyone starts becoming part machine—through bioengineering, through implants, through computers, through robotics, through all the press of modern technology—then the old "patriarchal dualisms" should start to disappear. If "man" and "woman" start fading away as natural categories, then sexism may vanish, too. "The cyborg is resolutely committed to partiality, irony, intimacy, and perversity," she wrote. "The cyborg incarnation is outside salvation history. Nor does it mark time on an oedipal calendar." Cyborgs, she wrote, "might consider the partial, fluid, sometimes aspect of sex and sexual embodiment. Gender might not be global identity after all, even if it has profound historical breadth and depth."[106]

Such thinking mostly appealed to avant-garde academics who were suspicious of the idea of solidity anyhow; this was the very heyday of poststructuralist literary criticism, the idea that "texts" had no fixed meaning. But Haraway's point made a certain sense. Perhaps we hadn't chosen to be modified, and perhaps it might lead to new kinds of domination, but it also might prove liberating: we weren't *just* men and women anymore, or black and white, or whatever—we were also part something new, and that stirred up new political possibilities. Just like the rhetoric about choice, the notion packs a certain deep appeal.

But in the end, it's mostly an academic argument—a game, by definition, played for pretty low stakes. Cyborgology gave up on the old ways of building a just society (the bumper stickers, the endless meetings with the Congressional Black Caucus) and substituted a technological end run. Even if, in the short term, that proved liberating, it was a freedom unlikely to last more than a single generation. For in its wake came not children who were creating their identity from the pieces they found around them, but children whose identity had been created for them. Not people cleverly mimicking robots, but people displaced by them. (Even the proponents of human genetic engineering, after all, admit almost casually that it will produce an elite GenRich unable even to mate with the rest of humanity.) These technologies shake up meaning *too much*. The "patriarchal dualisms" might erode away, but so would pretty much everything else. In the end, genetic technologies are no more a route to justice than a route to liberty.

All the proponents of the new technologies agree on just one thing: who their enemies are. Or, at least, who they'd like them to be: right-wing religious fundamentalists, the kind of people who don't believe we descended from monkeys, offshoots of the anti-abortion movement. According to Robert Weinberg, of MIT, all the clashes about, say, "therapeutic" cloning "converge on a single question: When does human life begin?"[107] Speaking at the UCLA conference on germline engineering, the ethicist John Fletcher said, "Religion plays a very conservative role in its response to genetics. . . . In my experience, very few deeply religious people are open to understanding biological evolution."[108] If you go to the Web, you'll find the technotopian chat rooms ringing with assertions like "Bush's declaration of the 'immorality' of cloning is like the shrieking call to jihad of the American Taliban," and "the anti-tech eco-terrorists and power-addicted PC ideologues of the extreme left are the American equivalent to the hate-filled, dogma-driven, anti-west, anti-modern, jihad psycho-killers of Al Qaeda."[109] In the liberal *American Prospect* magazine, stem-cell champion Chris

Mooney simply and memorably referred to opponents of cloning as "the embryo-worshipping religious right."[110]

They should be so lucky. In fact, as the first glimmerings of what this new future means began to seep into the public consciousness, all kinds of odd new alliances started forming. A few people thrill to this new posthuman vision; but more people, it turns out, feel a kind of gut revulsion. All kinds of people.

Feminists, for instance. Judy Norsigian, a member of the Boston Women's Health Collective and the current editor of *Our Bodies, Ourselves,* organized a letter opposing human cloning, signed by over a hundred pro-choice and women's health leaders.[111] "To describe her as pro-choice would be akin to describing the pope as Roman Catholic," wrote one journalist.[112] And yet, on the issue of cloning, which Norsigian called "very much a woman's issue," the *New York Times* described her as "comfortably aligned with conservatives."[113] When she testified before Congress, one Colorado Democrat, Diana DeGette, a supporter of therapeutic cloning, told her, "Your book is one of the most important in my life." The congresswoman's face, according to the *Chicago Tribune,* was "a study in cognitive dissonance."[114]

Some environmentalists have broken ranks with some parts of organized science, their usual allies in the fight against climate change or for habitat conservation. "The idea of redesigning humans and animals to suit the primarily commercial goals of a limited number of individuals is fundamentally at odds with the principle of respect for nature," Brent Blackwelder, the director of Friends of the Earth, said as the first cloning debates began. John Passacantando, the director of Greenpeace USA, signed on in opposition to cloning, as did the director of the Earth Island Institute. "We cannot mistake fundamentalism as the only enemy," Carl Pope told the annual convention of the National Abortion and Reproductive Rights Action League. "We must be as vigilant for the whirlpools of nihilism and extreme instrumentalism as we are for . . . Operation Rescue."[115]

Weird coalitions began to coalesce. Not all conservative politicians turned out to be anticloning; those with a strong libertarian bent or with a lot of high-tech industry in their districts were wary of strong restrictions on embryo research. On the other hand, Bernie Sanders, the Vermont independent and self-described socialist, voted with Tom DeLay, the leader of House conservatives, for a stiff cloning ban; so did eight other Democrats with 100 percent pro-choice voting records.[116] Petitions circulated signed by people like William Kristol, the editor of the conservative *Weekly Standard,* but also by people like Tom Hayden and Todd Gitlin, stalwarts of the sixties left. "I think a lot of people are looking for a place to stand which, in the highest sense of conservatism, preserves some sphere of life," said Gitlin.[117] Meanwhile, Sam Brownback, the Kansas senator and archconservative, was on *Meet the Press* sounding like Che in his concern for "the commodification of the human species."[118]

These are, in other words, very odd issues. And they will get odder as they get larger—when we move on from therapeutic cloning to actual germline engineering, or have to decide whether or not to fill the atmosphere with law-enforcement nanobots.

Perhaps we will simply give up and conclude that the question is too complicated, that "letting the market decide" is all we can hope for. But perhaps not. It's possible to imagine a politics emerging that takes technology seriously, that begins to wrestle with it as intelligently as, say, the Amish do, though on very different terms. A politics that coalesces, though not explicitly, around some version of what I've been calling the enough point. A politics that over time generates the net of regulations, and hence of taboos, that keeps us more or less human. We'll never win a permanent victory over these technologies—just as you can only save a wilderness area one year at a time, just as the strongest treaty won't make physicists forget how to build nukes—so germline engineering will always be out there, tempting us. A new part of what it means to be human is to live with these possibilities, and to guide, direct, and, when necessary, foreclose them. Francis Fukuyama has been especially levelheaded in his discussion of how all this might happen. "There are a huge number of permutations and combinations of possible rules that societies can establish," he says. They won't work perfectly—"no regulatory regime is ever fully leakproof"—but then, people still rob banks.[119] The key is to ensure that the robberies don't happen so often that they make banking impossible.

These new rules would require that we think more carefully than we have in the recent past. Instead of the constant glorification of the magic of the market, for instance, we'd have to start asking what corners of our lives are too precious for it to invade. We'd have to start considering more carefully what we owe to society (which is to say, what we owe to children in general, and to the future) as distinguished from what we owe to our own individual children in our own particular moment. We'd have to confront some hard truths—for instance, that we've let a proper concern for individual rights turn into a hyperindividualism that endangers the very species.

The new technologies are large enough to raise those fundamental questions. Not head-on, of course—our senators are unlikely to engage in robust debate about individualism and its limits—but in one specific decision after another. Should preimplantation genetic diagnosis be allowed for all diseases, or only for life-threatening ones? How much do we need to restrain embryonic stem cell research in order to ward off human cloning? Taken question by question, this politics will over time yield a working definition of enough. Over time, they'll let us say, "This far and no farther." They'll lead us bit by bit toward the revolutionary idea that we've grown about as powerful as it's wise to grow; that the rush of technological innovation that's marked the last five hundred years can finally slow, and spread out to water the whole delta of human possibility.

But those decisions will only emerge if people understand this time for what it is: the moment when we stand precariously on the sharp ridge between the human past and the posthuman future, the moment when meaning might evaporate in a tangle of genes or chips. As we've seen, human meaning turns out to be fragile—we can either pile sandbags around it to keep it safe, or watch it wash away.

And if it goes, it will take democracy with it. Forever. Our parents and grandparents fought the good fight against Hitler. Had he won, our democracy would have been wrecked. We might not be voting, or writing books, or talking freely. But Hitler's triumph would not have been *permanent;* even he only spoke of a thousand-year Reich. Just as dissidents across Eastern Europe kept alive a flickering hope of democracy throughout the Soviet era, some here would surely have done the same in a Nazified West. Genetic engineering, though, would be a different kind of attack; the "soft dehumanization" that comes with these new technologies might well quell whatever spirit it is that animates our life together.[120] If you didn't know how *you* felt, or if you felt how you did because your rejiggered cells were pumping out designer proteins—in that kind of world what, if anything, would "democracy" mean? If you're engineered for optimism, or for piety, or for trust? Democracy depends on the idea that we're free actors. Yes, class and prejudice warp us, and in turn that's warped our democracy. But not irrevocably, because those things are not irrevocable. At our best we've overcome them.

These new technologies are not yet inevitable. But if they blossom fully into being, freedom may irrevocably perish. This is a fight not only for the meaning of our individual lives, but for the meaning of our life together.

CHAPTER FIVE

Enough

In early August 1999, the fourth national convention of the Extropian movement convened in a California conference hall. In some ways the group was a marginal fringe of the most zealous technotopians—but its board of directors included Marvin Minsky and Ray Kurzweil, and the keynote speakers at its very first conference were Hans Moravec and Eric Drexler; in the course of that 1999 meeting Gregory Stock gave a talk on aging research, and Roy Walford described his experiments with caloric restriction.[1] The Extropians were, in other words, a wired outfit, and had in fact appeared regularly and prominently in the pages of *Wired*.[2]

In the evening, a man named Max More stepped to the podium in that California conference hall. He'd been christened Max O'Connor, but, just as he'd coined Extropy to be the opposite of Entropy, so he'd picked his new name as a sign of commitment to "what my goal is: always to improve, never to be static. I was going to get better at everything, become smarter, fitter, and healthier. It would be a constant reminder to keep moving forward."[3]

And then he delivered a talk entitled "The Ultrahuman Revolution: Amendments to the Human Constitution." It took the form of a letter to Mother Nature, and began by offering brief thanks to her for "raising us from simple self-replicating chemicals to quadrillion-celled animals."[4]

Graciously granting that Mother Nature had done her best (indeed, as More later noted, he was himself a stud, in the gym five days a week. "I can boast that I do 710 pounds on the leg press. No atrophied body here!"),[5] he went on to list, however, the "many ways you have done a poor job with the human constitution." He complained about the unfairness of dying "just as we are beginning to acquire some wisdom" and noted that we'd been given limited senses, imperfect memories, and "poor impulse control." Not only that, but Mother Nature "forgot to give us the operating manual."

As a result, More continued, "we have decided that it is time to amend the human constitution." In the course of the evening's talk he proposed seven amendments, invoking all the technologies, and all the aspirations, we've encountered in these pages. The elimination of death in favor of eternal life topped his list, followed by increased perception through "novel" and improved senses, enhanced memory and intelligence, improved "emotional responses," and reshaped "motivational patterns." Such gains would come "initially through biotechnology" and later through integrating other advanced technologies into our bodies; in every case, he declared, "we will not limit our emotional and intellectual capacities by remaining purely biologic organisms." In other words, he concluded, his menu of enhancements "will move us from a human to an ultrahuman condition."[6]

According to the official record of the talk on the Extropy Web site, he sat down to "sustained applause."

As well he should have, having expressed more succinctly than anyone before him the technozealot's creed. More listed all its particulars—and, to one extent or another, every item on his list has already been accomplished in the lab with other animals. We have worms living seven times as long as usual, and mice running mazes twice as fast. He's talking big, but not impossible. He also captured the basic dogma: that *human beings simply must push on.* Forget all the practical arguments why this work is inevitable—forget the difficulties of surveillance, the cheapness of the equipment, the lure of big money. At bottom, advocates insist, it's inevitable because human beings inevitably move forward, expanding their powers. In 1492, Columbus sailed the ocean blue; in 1969 Neil Armstrong took "one giant leap for mankind"; and sometime very soon there will be a baby born with improved hardware. By our nature we must crack the nucleus of the cell. From the human, we jump to the "ultrahuman" and someday, doubtless, to the doublesuperultrahuman.

Like Columbus sailing west, we have only the vaguest notions of where we might be heading. An "unboosted human brain" could never have a real conversation with one of the coming immortals, writes Damien Broderick, could never know "what vast issues" it was considering.[7] Oh, we can guess at the wonders, just as Columbus anticipated spice and gold. One of More's colleagues, a Swedish philosopher named Nick Bostrom, took to the podium at the same conference to predict "orgasms and aesthetic-contemplative pleasures whose blissfulness vastly exceeds what any human has yet experienced," and "love that is stronger, purer, and more secure than any human has yet harbored," and "values that will strike us as being of a far higher order than those we can realize as unenhanced biological humans."[8] But the destination hardly matters; what's important is the trajectory, the surge, the momentum. Forever upward, forever more, forever restless. That's the reason, in the Extropian view, that we have minds: to push forever ahead, transforming ourselves ever and again into something new.

Such a vision of who we are obviously preempts the very idea of enough. If the technological visionaries understand human beings correctly, then the human spirit will, with uncheckable dynamism, simply overwhelm whatever rules and taboos some of us might propose. If our destiny lies ever further on, always just out of our grasp—well, who are we to argue with destiny? *Homo sapiens* will be left behind on the accelerating curve of progress, and our descendants will be off to the stars, or the computer banks, or some other place too complicated for us to understand in our current primitive state.

That sense of man as constantly, inescapably, and quintessentially a striver, a builder, an engineer, a creator is deep and powerful. In the early days of the Enlightenment, the Marquis de Condorcet stirred Europe when he declared "the perfectibility of man is absolutely indefinite . . . the progress of this perfectibility henceforth above the control of every power that would impede it, has no other limit than the duration of the globe upon which nature has placed us."[9] Our intelligence, wrote the philosopher Henri Bergson, "is the faculty of manufacturing artificial objects, especially tools to make tools, and of indefinitely varying the manufacture."[10] We are cleverer beavers. As often, Gregory Stock puts the matter most directly: "To turn away from germline selection and modification without even exploring them would be to deny our essential nature and perhaps our destiny. Ultimately, such a retreat might deaden the human spirit of exploration, taming and diminishing us." Now that the continents are filled, "exploring human biology and facing the truths we uncover in the process will be the most gripping adventure in all our history."[11] We must push forward. We have no choice. Enough is not a possibility for our species.

If all this sounds grandiose, it's in fact just the opposite. The reason the technotopians can talk so casually about the "posthuman" future is that they find nothing particularly significant about the human present. According to them, we engage in this constant push forward not because we're so high-minded or passionate or special, but because we're not special at all. Because we *literally* have no choice. Nothing about us sets us apart from other organisms. Our bodies are "nothing more than bio-molecules interacting."[12] Our brains, in the words of Marvin Minsky, are "meat machines."[13] As Dr. Robert Haynes, president of the Sixteenth International

Congress of Genetics, told his organization, "For at least three thousand years, the majority of people have considered that human beings were special. . . . What the ability to manipulate genes should indicate to people is the very deep extent to which we are biological machines. . . . It's no longer possible to live by the idea that there is something special, unique, or even sacred about living organisms."[14] This is no small point. Provided you believe it, you can stop worrying about human meaning disappearing, because it wasn't really there to begin with.

Rodney Brooks titled one chapter of his book on the fabulous future "We Are Not Special." Molecular biology, he writes, "has made fantastic strides over the fifty years, and its goal is to explain all the peculiarities and details of life in terms of molecular structures. A central tenet of molecular biology is that *that is all there is.*" These molecules interact with each other according to "well-defined laws," combine in predictable ways, producing in our case a body that "is a machine that acts according to a set of specifiable rules. . . . We are machines, as are our spouses, our children, and our dogs." And now we are building machines that will match and surpass us. "Resistance is futile."[15]

This logic should appeal in certain ways to environmentalists, who tend to agree that man has made too much of himself, that we've indeed valued our own species too highly. And often both engineers and environmentalists are especially harsh in blaming religion for that pride; they take a certain savage pleasure in demolishing its claims. They trace their lineage back to Copernicus and Galileo, who with new observation and instrumentation shattered the idea that the universe revolved around the earth—shattered the idea that the earth was special when in fact it was "off to one side in a galaxy of billions of suns."[16] On earth, however, man was still set apart, the lord of dominion, specially created by God in his image. Until, of course, Darwin shattered that conception as well—we weren't special, we were, in the words of Gregory Paul and Earl Cox, just "souped-up apes, produced by achingly slow, chaotic, Rube Goldberg evolution over billions and billions of years of countless generations of death, extinction, waste, and suffering. Awfully slow and clumsy for an omnipotent deity."[17] When Crick and Watson elucidated the structure of DNA, "there was a further weakening of specialness. Soon it was found that all living creatures shared the same sort of DNA molecules and used the same coding scheme. . . . Even worse, we humans shared regulatory genes, relatively unchanged, with animals as simple as flies. There were even close relationships between some genes in humans and in yeast." These discoveries, says Brooks, "hammer home the nonspecialness of humans."[18]

That humans still believe in something "mystical" is an anachronism, these prophets say, one that will fade as we turn these new discoveries into technologies. "Who will need an eternal life-giving God when eternal life is available by alternative and real means?"[19] Eve and Prometheus and Pandora all shrank the domain of the gods—and now we shall do so again, finally, permanently. Whom *would* you worship as your creator if your genes came from Pfizer? If your daily bread came straight from a magic nanobox? If you had been programmed? Eventually, like all other meanings, religion would wither away. That's a lot of human legacy to dispense with, but we might well do it. According to the technotopians, we *will* do it. We have no choice; we inevitably push forward. It is our destiny, and destiny is inescapable. We can't be in control. *We aren't special.*

Except for one thing. Just one small thing, which the apostles of our technological future have overlooked. One small thing that actually does set us apart.

What makes us unique is that we can restrain ourselves. We can decide not to do something that we are able to do. We can set limits on our desires. We can say, "Enough."

This may not sound so striking at first, but it's the wild card in the deck, the joker that can save us yet, even though the technozealots are sure they're sitting on a winning hand. I'll close by exploring this idea, because I think it's our last best hope to prevent the wholesale loss of meaning that we now face.

And I'll begin with the beavers that live behind my house, the beavers I hear slapping almost every night on the marsh they've built in the aptly named Beaver Brook. Beavers, slapping their tails against the threat of passing canoes, possess a goofy charm; they're about my favorite animal. Still, there is something remarkably compulsive about them. One year a family built a small lodge on our pond, and the male swam across each evening on his way to work. He was as regular as a Swiss train, five o'clock each afternoon; you expected to see him carrying a lunch bucket. He and his crew build dams. They need to, in order to make sure that the holes to their lodges stay safely below water. And they need to gnaw on trees, or else their teeth will keep growing until their mouths are wedged permanently open. These needs have turned into what we call instincts. Strong instincts. If you want to see a beaver, here's all it takes: sneak out to a dam and pull a couple of logs out (easier said than done—beavers are remarkable builders). The sound of water trickling down the dam will, within a very few minutes, bring them from their dens. They *need* to stanch that flow.

Now, we all have that beavering drive within us. (In fact, we have it in spades. Beavers content themselves with one dam at a time. No beaver has *chains* of dams. They don't *franchise* dams.) When the engineers say that we are driven constantly to surmount any limit, we know what they're talking about. Robert Frost once wrote a poem about a man who plants a peach tree outside his New England home, and then spends the coldest night of the year wondering whether it is surviving the chill. "What comes over a man, is it soul or mind / That to no limits and bounds he can stay confined?" asks Frost. Only the advent of spring will tell if the tree has survived, "But if it is destined never again to grow, / It can blame this limitless trait in the hearts of men."[20] This "limitless trait" has led each of us to both glory and shame; it is integral. Even the saints feel it; indeed, in their struggles they feel it more than most of us.

But to say that we wish to surpass limits does not describe us fully. We are also the creature that can say no. The creature that, in Erazim Kohák's lovely phrase, can "subordinate greed to love."[21]

Take dams, for instance. We build them too, obviously—build them higher than beavers do, build them stronger. I've seen the largest earthen dam on the globe, built by Hydro Quebec on the La Grande River near Hudson Bay: a dam so mighty that its spillway could carry the combined flow of all the rivers of Europe. On the other hand, we also don't build dams. The modern environmental movement got its start when John Muir formed the Sierra Club to battle the dam planned for a California canyon called Hetch Hetchy. He lost that fight, but in the process saved Yosemite, just as David Brower, the great twentieth-century American environmentalist, saved the Grand Canyon from a plan to plug the Colorado. They were able to rally people by appealing to the *other* parts of our nature, the parts that aren't always striving and questing and grasping. Not the limitless parts, but the limiting parts. The parts that understand beauty and scale, the parts that sympathize with the rest of creation, the parts that can imagine sufficiency. Hydro Quebec built that huge dam on the La Grande, but so far the company's plans for an even bigger dam have been stalled; across North America, people concerned about the rights of the Cree Indians, about the caribou, about the sheer existence of a vast wilderness, have scrapped and battled to rein in the project. In 1999, then–U.S. interior secretary Bruce Babbitt pushed the plunger to dynamite the Edwards Dam on the Kennebec River in Maine, the first operating hydroelectric dam in the nation ever to be intentionally destroyed, in this case to make way for fish. "This is a statement about our capacity to honor and respect God's creation, the sacramental commons, and to live not just in the past, but in a visionary and different future, in a way of harmony and balance with creation," said Babbitt.[22] And it wasn't the only such sign. Elizabeth Grossman, in her recent book *Watershed,* shows that across the country "the pace of dam removal has overtaken the pace of construction as communities commit themselves to river restoration."[23]

Human beings, in other words, can be more complex than the engineers give them credit for. Or the economists, who tend to believe, like Lester Thurow, that "man is an acquisitive animal whose wants cannot be satiated. This is not a matter of advertising and conditioning, but a basic fact of existence."[24] In order to keep this dogma intact, economists will contend that, say, breaching a dam satisfies other "wants"—the want of beauty, perhaps. They try, charmingly, to quantify the "existence value" of a free-running river by asking people how much they'd pay for such a wonder. But when they engage in those exercises, they admit that our minds and hearts are more complicated than a mere collection of wants. Yes, faced with starvation, most of us would cross most limits. But it doesn't take great wealth before the more complicated parts of our nature assert themselves: I think often of those Bangladeshi peasants in their organic zone, having chosen community and health over the newest seeds.

It's this ability to limit ourselves—in Kohák's words, "the recognition that something may be perfectly understandable and yet be *wrong*"—that makes us unique among the animals.[25] Not *better.* But *unique,* as birds with their hollow bones are unique, and dogs with their astonishing sense of smell. You could argue that the rest of creation manages to observe these limits with enormous elegance—spontaneously, without even trying. As for us, we are the creature that can voluntarily rein itself in. We are, in some sense, the sum of our limits.

And though it galls the apostles of technology, this idea of restraint comes in large measure from our religious heritage. Not the religious heritage of literalism and fundamentalism and pie-in-the-sky-when-you-die. The scientists may have drowned the miracle-working sky gods with their five-century flood of data. Copernicus and Darwin *did* deprive us of our exalted place in the universe. But this older, deeper, more integral religious idea survives. Indeed, it thrives whenever man is knocked from his pedestal, for at its core is the notion that meaning matters more than size, that we are great precisely as we are able to make ourselves small. It is Yama, the King of Death, explaining in the Upanishads the choice between *preya,* that which is pleasant, and *shreya,* that which is beneficial. It is Gilgamesh, the great hero, reminded that immortality is not for man. It is Job, finally silent and satisfied before God and the splendor of creation. It is Jesus, tempted in the desert by the nanotechnologist of his day: "If you are the son of God, command these stones to turn into bread." And refusing, in words that still carry a charge: "Man does not live by bread alone, but by every word that proceedeth out of the mouth of God."[26]

In this long tradition, meaning counts, more than ability or achievement or accumulation. Indeed, meaning counts more than life. From this perspective, Christ's resurrection is almost unnecessary: it is his willingness to die, to impose the deepest limit on himself for the sake of others, that matters.

The entrepreneurs of the germline and the robot future have found their house preachers, of course. Anna Foerst, who served as official theologian at the MIT AI lab, sees "engineering as prayer," and the new technologies as "a sign of people trying to participate in God's creativity."[27] Edmund Furse, a computer scientist and author of *Towards the First Catholic Robot?* urges that robots be baptized (though he adds that they might find church boring because they are capable of reciting the creed "a thousand times faster" than the other parishioners.)[28] The Chicago scientist and "serious Methodist" Richard Seed was one of the first to announce that he would set up a cloning lab: "God intended for man to become one with God," he said. "Cloning and the reprogramming of DNA are the first serious steps in becoming one with God."[29] Even Rael considers himself a spiritual leader.

But none of that touches the core of our religious understanding. In the Western tradition, the idea of limits goes right back to the start, to a God who made heaven and earth, beast and man, and then decided that it was all enough, and *stopped.* "And on the seventh day God ended his work which He had made."[30] At the time God declared that we are made in his image, all we really knew about him was that he thought the world was good, that he wanted us to take care of it, and that it was time to take a rest. We take that rest still: Sabbath, Shabbat is the weekly reminder of this other religious tradition. This faith has no formal creed, no official ritual, and yet it has persisted for millennia through its insistence that instead of putting ourselves at the center, we need to move a little to the side. The same tradition stretches back to the Buddha and runs up through Francis and Thoreau, the constant countercultural witness, the never-ending whisper in our ear that we'd be happier, more satisfied, if we laid aside our hopes for immortality, for power, for wealth. If we turned the other cheek.

This tradition has never disappeared, but it's never carried the day, either. Most of us mature only partway; we learn—it's to be hoped that we learn—to place our family or our community or our deity nearer the center of our lives, but only in rare cases do we fully vanquish that compulsive striving, that grasping for more. And in recent centuries we've come to embrace our selfishness—our hyperindividuality—with an almost religious fervor. A few epidemics of questioning have occasionally swept the land—the countercultural sixties, for instance—but they didn't last long, and were easily coopted. Sunday means football and shopping as much as it means rest. The choice between Enough and More has always been a choice we could put off a little longer, both in our own lives and in the life of our civilization.

But now the hour draws near. Faced with a challenge larger than any we've ever faced—the possibility that technology may replace humanity—we need to rally our innate ability to say no. We will be sorely tempted to engineer our kids, but it's a temptation that we need to resist as individuals, and to help each other resist as a society.

The choices that we face, in fact, will settle this question of specialness once and for all. If we cannot summon our self-restraint, or if it proves too weak, then we will leave our uniqueness behind forever. Once we start down the path of turning ourselves into machines, of writing ineradicable programs for our proteins, there will be no way, and no reason, to turn back. We'll do what our programming indicates, never knowing how much choice we really have. We'll be like obsessive-compulsives, for whom some accident of wiring or chemistry has overridden the ability to choose. They must behave, repeatedly, in some particular fashion that they usually realize does not reflect their true self. They feel as if they have no choice. But terrible as their condition is, it can yield to the liberating effects of reflection, therapy, medicine.

It won't be faulty wiring, though, that robs the engineered of their agency—it will be intentional programming. We'll do what we're supposed to do: we'll be brainy or brawny or pious. We may not feel sad—we won't necessarily want to be liberated from it—but we'll live in a world where our humanity really has vanished. The tensions, in other words, between our unlimited desires and our capacity for self-restraint would simply disappear. We'd be on the *more* track. If you're designed to be athletic, you will always choose more speed and power, and you'll choose them for your children. The tension between that athletic part of you and the other parts will simply disappear; you won't question yourself at mile 23 of the marathon. If you're designed for piety, the temptations of the world may barely arise. Because, of course, those tensions are inefficient, like feathers on a chicken. They keep us from being all one way, one thing. From specializing emotionally. But that inefficiency, that tension, that tug in different directions is what we call consciousness. It explains novel-writing and rock-climbing and church-going, and it explains both the difficulties and the glories of family and community and love. Machines don't have that tension; the other animals move along that edge with inborn grace. Consciousness doesn't make us better than robots and rhinoceri. It just makes us different. It just makes us human.

The idea that by escaping the body we will become "everything" accords very nicely with the economic worldview that we can never be sated, with the scientific paradigm of eternal progress. But in the back of our heads a much older wisdom whispers that should we ever escape our limits we will become—nothing.

"Limits" sounds so negative. So unpleasant. Eat your bran. There is something extremely seductive about this notion of going on forever forward, of never saying, "Enough." It's dynamic! We'll be smarter, fitter, healthier. We'll press 840 pounds on the leg machine! We'll see in six dimensions. We'll have eyes all over our heads. We'll have a box that cranks out anything we want. We'll live forever. Pass the ice cream.

But it's not just religious tradition that suggests that satisfaction comes from honoring limits. The arts—the record of what we have found beautiful or meaningful or both—make the same claim. Music and painting and dance and writing depend on certain formal limits for their sense, lest they become mere notes, mere nonsense syllables. In a deeper sense writer Robert Pack makes the case that great art depends on the acceptance of what Yeats once called "life at peace with itself." Artists have always had the power of gene splicers or nanotechnologists: they could exercise enormous control over the lives they created. "All artistic creation partakes of the wish for power and control, which in its fullest fantasy is the wish for immortality, to exist beyond the bonds of nature," Pack writes. "But such a wish, persistent as it has been throughout human history, is doomed to failure. We must return to our lives, to the limits of our aging bodies."[31] We don't want to; indeed, artists often fight hardest and most violently against happiness, because "affirmed pleasure brings us to a limit, forces the awareness of that limit, and thus reminds us of our mortality."[32] "I can't be any happier than this" is a phrase with two meanings.

But if we can't accept that we will die, there can be no harmony, only the fantasy of an afterlife or the illusion of a kind of immortality through art. If, says Pack, the artist can't make peace with the idea that she owes the world a death, that she will leave room for those who follow her, then she will "have no theme but her unhappiness."[33] Macbeth is our archetype of the man whose ambition leads him to reject the natural order, in his case by killing the king. The consequence is that he loses all that is meaningful in life—the blessing of sleep, trust in others, the comfort of family, the power of compassion. Even, most tellingly, "his sense of the seriousness of life, making of his own ambition an ironic mockery."[34] In the end, life becomes literally meaningless, "a tale, told by an idiot, full of sound and fury, signifying nothing."[35]

Such limits may no longer be mandatory—we might, for example, be able to live pretty much forever and come to possess unchecked power over matter. We might turn stones to bread, or at least dirt to potatoes. In that case, the artist might take on "any" theme, not feel confined by the arc of our lives. But what, exactly, art would then consist of is hard to imagine. Maybe more artists would stick phosphorescent genes into rabbits. Or hamsters! Or babies! We'd have to jettison the art of the past, which was primarily the record of humans coming to terms with their mortality. The record of our attempts and failures to mature, and of the evanescent glories of life that does not last forever, and of the consolations that make it bearable. All that we would discard—it might be of historical interest, but not of human interest. It wouldn't bear on "life" in that world without limits.

Art would wither. Oddly, science would, too. Or much of it, the part that finds its drive in "mere" curiosity about the rest of life. Sure, engineers would have a brief heyday in this limitless new world. But what about the passionate observers? Leonard Hayflick, the gerontologist who discovered the natural limits of cell division, said he had no desire to manipulate the human life span. "Why, then, have I spent my lifetime learning about aging and longevity? . . . Because I am curious about how the world works."[36] But how would that curiosity survive very far into the future? It's not just that, as we saw in chapter 2, more "efficient" robot biologists would take over the work. It's that—who would care? In a world without limits, the only kind of "real" science would involve engineering the next jump up the ladder. Any research that bore on the project of human acceleration we'd be pressing ahead with; everything else would be pointless. Ecology, the great emergent science of the twentieth century, has tried to understand the incredible complexity of relationships that govern the living earth—tried to understand, among other things, how we might fit in. But if we finally and forever breach all those relationships, emerge as a limitless force, then what possible interest will any of the rest of life hold?

I first felt this kind of anticipatory sadness for a vanished science and art while watching Errol Morris's witty documentary *Fast, Cheap & out of Control.* It focused on four eccentrics—a lion tamer, a topiary gardener clipping his hedges into giraffes, a man obsessed with the study of a rodent called the naked mole rat, and . . . Rodney Brooks, the robotics engineer. In this company, he was as charming as the rest, one more guy following the odd dictates of his heart. But if *his* vision comes to pass, the world will be utterly and forever changed: the topiary gardener doesn't want all the rest of us clipping our hedges, but Brooks wants—or, at least, is willing to countenance—our replacement by "superior" machines. And machines, by definition, are ever less quirky. Their bugs get ironed out. The programming gets better. It's hard to imagine that the engineered creatures who follow us will be eager for even the kind of attenuated relationship with the wild that lion taming represents (this echo of deep human history, when lions meant something to us). It's difficult to conceive that these advanced, hyperrational creatures will be spending a lot of time on topiary.

At the moment, all of these passions can coexist. There is no danger that a fascination with the biology of mole rats will make the world unsafe for hedges shaped like giraffes. But if Brooks pursues his quest to the end—well, it's not just an eccentric and charming notion. It's one that erases everyone else's vision. Consider paddling around a lake in a canoe: fifty canoes can be exploring the bays and coves and not bothering one another at all. But one motorboat roaring through changes everything for everyone. It may be grand fun for the one guy standing at the wheel with the wind in his hair, but everyone else is left to deal with the wake and the racket and the diesel stink as well as they can.

A technological enthusiast named John Brockman recently set up a Web site, edge.org, that brings together many of high tech's leading lights for a series of virtual symposia. On the occasion of the millennium, he asked his digerati to name what they considered the most important invention of the last two thousand years and explain why. The results outline with wonderful precision the full-steam-ahead worldview, the sense that what counts about us is our ability to smash through limits.

Many of the best and the brightest who responded were clearly out to demonstrate their cleverness: they nominated the use of hay as an animal feed (which let society move steadily north from the Mediterranean), anesthesia, Gödel's incompleteness theorem, the patent office, the eraser, double-entry accounting, the Gatling gun, and "the nonimplemented 33-year English Protestant calendar."

But most of the answers converged on a relatively few inventions, and on an even more central concept: what mattered most were *catalysts,* those inventions that speeded up the process of invention. John McCarthy, a pioneer in artificial intelligence, said, for instance, that "the most important invention is the idea of continued scientific and technological progress," an idea "institutionalized with the patent laws." The less sweeping answers sounded the same theme. Several people submitted the invention of calculus, or the zero, or the Hindu-Arabic number system: "Without it," explained the mathematician Keith Devlin, "Galileo would have been unable to begin the quantificational study of nature that we now call science."

Great numbers of respondents chose the computer—"because of the way it extends the capacities of the human mind," because it "will govern everything we do in the next twenty centuries," because it will be the springboard for the distribution of "life and intelligence . . . across the cosmos." Inventions that seem at first glance unrelated were nominated solely for being predecessors of the computer: paper, according to the IBM researcher Clifford Pickover, because "it's a progenitor of the Internet." Electricity, says the science writer Margaret Wertheim, because "it is the ability to transport electric power at the micro level that has made possible silicon chips, and the attendant computer and information revolution." Charles Simonyi, the chief architect at Microsoft, nominates "public key cryptosystems" as the most important invention since the birth of Christ because they let you use the Web without giving up your privacy. And, again and again, the printing press. Not for the books that rolled off it, mind you, but because it led to the Web. "Gutenberg would be pleased to see where his invention has taken us," declares Lew Tucker, a "Java evangelist" who is the director of developer relations at Sun Microsystems.

All in all, virtually everyone who responded proposed some invention that mattered because it made the next set of inventions easier. Even the writer Douglas Rushkoff, who nominated the eraser, did so because it lets us get rid of our mistakes and make faster forward progress. (One guy voted for the symphony orchestra on the grounds that its spirit of cooperation was "a symbol for something that may be yet to come, like space travel.") A single example of what you could call a limiting technology was nominated: the contraceptive pill, which began the revolution in human fertility that has allowed people to bring population growth under some control. But the man who proposed it, Colin Blakemore, the chairman of the chief decision-making body of the British Association for the Advancement of Science, said its importance lay elsewhere: it "triggered a cultural and cognitive revolution in our self-perception. It has contributed to our ability to accept organ transplantation, the notion of machine intelligence, gene therapy, and even, eventually, germ-line genetic manipulation."[37] Forward, ever forward.

Given the assignment, no one can argue with the picks. These inventions *have* changed the world. But they say more about the last fifty years and the last five hundred than they do about the last two millennia. They are, in other words, the product of a particular and impatient moment, a particular way of seeing the world. The way that leads straight off the various cliffs this book has described.

But the other path I described, the one that embraces our limits, also runs through our time. It may be less well traveled, but it's not completely overgrown with tall grass, either. We've battled nuclear weapons with resourcefulness and success, begun the work of cleaning our air and water. In fact, there's a whole different set of inventions from our time that point toward enough, resources from our recent past on which to draw. Consider just two innovations of the twentieth century.

First, the invention of nonviolent civil disobedience as a method for political change. Gandhi's new technique freed the second-largest nation on earth from the control of the most powerful empire the world has ever seen, without a war. He taught people in large numbers to do what Jesus had recommended to individuals: turn the other cheek. He understood that nonviolence was not only a moral, aesthetic stand, but a potent political technique as well. One that was pursued by Martin Luther King, Jr., with notable results, and one that, if further developed, could make the world a stunningly different spot. It is no coincidence that Gandhi was also the most powerful twentieth-century spokesman for the proposition that less is more, that human satisfaction lies in respecting material limits, in opening yourself to the claims of others, in backing away from the hyperindividualism of the West. Nonviolence was a method of limiting oneself that inspired others and held back adversaries—a remarkable technology.

Second, the invention of the wilderness area. My backyard, the Adirondack Mountains of northern New York, was set aside by the state legislature in the 1890s as "forever wild"; not a tree was to be cut. Seventy years later, the U.S. Senate did the same thing on a larger scale, designating wildernesses across the nation to remain "untrammeled by man." Now "biosphere reserves" with large wildernesses at their core have sprouted in countries around the world. They are crucial ecologically, but in some ways their greatest value is philosophic: here are places where people have actually decided to take a step back. Where they've decided that other species, other needs, are more important than ours.

Nonviolence, wilderness—these are the opposite of catalysts. They're technologies that act as brakes, that retard our pell-mell rush forward, that set sharp boundaries on where we're going and how we'll get there. Right now, they aren't as important as computers. But one can at least envision a world in which they might be. We've not yet foreclosed that planet; enough remains a possible invention.

One of edge.org's deep thinkers, the Amherst College biologist Paul Ewald, offered "the concept of evolution" as his candidate for the most important invention. "It offers the best explanation for what we are, where we came from, and the nature of life in the rest of the universe," he wrote. "It also explains why we invent, and why we believe the inventions described in this list are important. It is the invention that explains invention."[38] Ultimately, Darwin pushes us onward: we keep accelerating because "evolution" leaves us no choice. That idea exerts enormous power. It's the trump card in the winning hand the techno-utopians think they're holding. "Why did I splice genes? My genes made me do it." It sounds unbeatable

But recognize first that the claim is not meant to be taken literally. Nobody actually believes that "evolution" as explained in *The Origin of Species* drives us to germline genetic engineering or nanotechnology. That is, if these technologies are developed, it won't be because people who wanted to splice genes were more reproductively fit, and hence over thousands of generations their offspring came to outnumber the offspring of the cautious, so that the odds increased that we would start adding additional eyes to our children. Such classic biological evolution takes place very slowly, and in the human case may have essentially stopped: medicine (and fertility clinics) means that most of us, no matter what our temperament, reach reproductive age healthy enough to bear children.

Instead, the techno-utopians use evolution as a metaphor. "In the same way" that fish came out of the sea and learned to walk, we are going to learn to augment our intelligence or beef up our muscles. For the fish, it happened slowly and "automatically." For us, it will happen quickly. And we will choose it, but through a nearly automatic process that these prophets sometimes call cultural evolution. At first this kind of evolution was nearly as slow as biology: "For tens of thousands of years, humans had created tools by sharpening one side of a stone," writes Ray Kurzweil. "It took our species tens of thousands of years to figure out that by sharpening both sides, the resultant sharp edge provided a far more useful tool."[39] But now cultural evolution is speeding up—indeed, says Hans Moravec, it is "a self-accelerating cycle that is reaching escape velocity today."[40] This new kind of "unnatural selection," as Marvin Minsky calls it, works even better than the old kind. Why? "Because we can exploit explicit plans and goals."[41] Unlike evolution, which is notorious for having no end in mind, "we may choose one for ourselves."[42]

Something like cultural evolution clearly exists. People figure out some useful trick—fire, say—and pass it on to their neighbors; they don't have to wait for their genes to do the job through natural selection. But to claim that cultural evolution *compels* us is a fishy little intellectual bait-and-switch. On one hand, we have evolution—implacable, impersonal, without a goal. On the other hand, we have human beings making choices about their future. You don't get to have both. Either it's a choice—*in which case we could choose not to do it*—or it's a force that simply can't be denied. In other words, the metaphor is wrong.

This gives us a chance to employ other metaphors that point in different directions. For instance, perhaps we should stop thinking of human beings as a species. As I mentioned, we're not really "evolving" like a species anymore. Instead, substitute another metaphor: the human species as one large individual organism.

Unlike species, which keep "evolving" forever, individual organisms follow a very different pattern. They grow for a while, and then they stop growing. This is the process that a biologist would call maturation. When you were young, your hormones spurred you to change, to grow. This was perfectly natural, perfectly healthy. If it hadn't happened, your parents would have taken you to the doctor, and she would have diagnosed you with "failure to thrive." We need to grow. But in late adolescence, that tide of hormones begins to recede and shift. Our bodies have reached full size. We no longer need to expand. In fact, if you are twenty-three and still adding two or three inches a year, you'll have to visit the doctor; she'll diagnose some pituitary problem and do all that she can to stop you from growing. You've reached a proper size.

By analogy, try to imagine the same thing happening with us as a species, as a society. Having grown very slowly for a long time, and then gone through a stupendous five-hundred-year growth spurt, we might at least consider the possibility that we are big enough. In that case, we would need to choose to stop; it won't happen automatically, any more than we are "automatically" forced to go forward into the posthuman world. We would need to pass the laws, form the committees, enforce the regulations to make sure we don't, say, engineer the germline.

We've done this with other problems—recognized signals from the world around us that showed we were getting too big and taken steps to make sure it didn't happen. When word of the population explosion filtered around the planet, most governments in the developing world began working to persuade their populations to reduce the number of children they had, and most developed countries chipped in to help pay for the effort. As we have seen, it's worked (a little too slowly; but then, many adolescents mature a little too slowly). Now, shocked by signs that the planet is warming at a dangerous pace, most nations of the earth have come together to try to cut down on the carbon we pour into the atmosphere. Again, the brakes are being applied too slowly, and in this instance with no help from the United States—but then, perhaps we're not the most mature of countries.

Right now, plenty of people feel the peacefulness of their lives degraded by sprawl, or worry about the way consumerism has eroded the quality of our communities. For them, the idea of enough is not completely alien or distasteful, though it remains difficult to embrace. We've been told that it's impossible—that some force like evolution drives us on to More and Faster and Bigger. "You can't stop progress." But that's not true. We could choose to mature. That could be the new trick we share with each other, a trick as revolutionary as fire. Or even the computer.

Maturation is not "stagnation." Organisms don't stagnate when they stop growing. People who reach twenty and stop getting taller don't start getting shorter the next day. Instead, if they are lucky, they begin to mature. That's always a bittersweet process; it involves letting go of the adolescent fantasy that you can do anything. It involves picking and choosing. Deciding what you're going to major in, what profession you'll enter, which man you'll marry. Those choices foreclose other options. By the time you're thirty, it's quite difficult to decide that you should have been a doctor. And that's sad. But not tragic. Maturing is as sweet as it is bitter. It involves, at heart, deciding that you're not the most important thing on earth, and that others—your community, your spouse, most obviously your children—are worth your devotion. The people we admire most are the ones who have most fully matured. They seem to us not only good people, but often satisfied and content. Fulfilled. And, not coincidentally, more likely to be able to deal with their eventual death.

Societies, species, could do the same. We could decide that instead of endless technological growth and economic expansion, we want to focus a larger fraction of our joint energies on other things: service, art, celebration, love. We are large enough and rich enough that such a decision is conceivable. It wouldn't require that we lead lives of sacrifice and poverty; merely that we lay aside childish fantasies of eternal wealth and eternal life. We could decide to stay mere humans, not to trade in what we have for the gaudy charms the technozealots are promoting.

In this metaphor, consciousness becomes a more subtle and complicated phenomenon. If we see ourselves as being inexorably driven forward by the irresistible forces of cultural "evolution," then consciousness is merely the nifty tool that we've hit on to speed up the process. It allows us to build better stuff and tell each other about it and ratchet everything up another notch. He's got a wheel; I'm going to get me one too. And mine's going to have a hubcap.

But maybe consciousness has another function. Maybe it is the gift—the specialness—that allows us to eventually put a *brake* on this kind of evolution. To slow ourselves down, to keep ourselves from driving down certain roads. The technologists talk about "memes," ideas like "the wheel" or "fire" or "freedom" that spread in the fashion of genes. But the awareness that "this kind of fire will burn you" is a meme, too; memes can be cautions as well as catalysts. By this light, our gut revulsion at the coming "enhanced" world is consciousness trying to save itself. As I've tried to show, the advent of these technologies, and this posthuman world, will quickly undermine consciousness. If we turn into engineered automatons, then consciousness as we know it—including the ability to make our own decisions, to say no—will eventually disappear. We will have reduced it to meaninglessness; human consciousness will have committed suicide. But if, on the other hand, our nay-saying ability proves strong enough to help limit our desires, and hence our technologies, consciousness may survive into the deep future. That's the immortality that should interest us, the "evolutionary goal" we should target.

Gregory Stock once wrote that "the human mind cannot be the highest summit of cognitive performance."[43] If we view cognitive performance in terms of how many computations the mind can do each second, or how much information it can instantly recall, that's clearly the case. But the human mind may nonetheless be the apex of thinking machinery, simply because it is able to hold things in balance. To weigh desires. To say no.

Saying no will never be easy. Maturing is the hardest work of our lives, and most of us make it only partway. I've been talking a brave game, but I don't know what would happen if someone knocked on the door this afternoon with a syringe full of whatever was going to let me live forever. Maybe I'd swallow the IQ pill, or stuff it down my daughter's throat. Maybe it's only the fact that I'm relatively well off, relatively healthy, and in my relatively early forties that lets me think I'd be able to say no, or to tell anyone else to.

Certainly that's how the proponents of the posthuman world see it. Once people realize that cryonics offers the real possibility for resurrection, says Robert Ettinger, "nearly everyone will see the Golden Age shimmering enchantingly in the distance and will not dream of relinquishing his ticket. . . . Before long, the objectors will include only a handful of eccentrics."[44] As soon as the genomics revolution really kicks in, writes Stock, "our agonizing about playing God" will "give way to a new chorus: 'When can I get a pill?' "[45]

Would we ever have enough strength to choose life as we've always known it, life that includes death? Always before, this passage has been smoothed for us by its inevitability: "Like death and taxes." "Everyone's gotta go someday." But if those things weren't precisely true anymore, how would we react? Especially given that we live in a world where much meaning has already eroded, where we don't have the support of, say, the medieval church, or even of a stable community in which we grew up and then raised our children, to help us see that there might be value in the steady rhythms of human generations. I can walk out my door and into the woods and hence into the endless repeating cycle of the natural world, which offers its own argument and its own consolation. Immortality matters less among the rotting trees and the sprouting saplings, just as "enhancement" matters less among people who take good care of each other. But fewer people in each generation can find such places to plant their feet. Each time some meaning, some context, disappears, it becomes a little harder to figure out a reason to hang on to what remains.

And yet, and yet. People are enormously strong. We *can* deal with death—it fractures very few of us. Not just Socrates, but ordinary people, every hour of every day in every town on earth, lay down their lives with as much grace as they can muster. A lot of grace, sometimes. To have seen people dying good deaths, people for whom the last months have been a time to sum up and knit together the lessons of their life—well, it doesn't make death less sad, but it does make death less scary. Some of us have watched our own parents go, and seen that the fact of their grandchildren playing at the foot of their deathbeds made it somehow okay and right. Life at peace with itself.

That is the choice that we, or our children, may be called on to make. A choice no human has ever made before, and one that no human should have to make. To give up our citizenship in the land of the finite, which is the place that humans have known, and trade it for a passport to the infinite. To be launched into a future without bounds, where meaning may evaporate. To live always in the future, and never in the now where humans have mostly dwelt. When Miranda exults in *The Tempest,* "O brave new world that has such people in it!" her father replies: " 'Tis new to thee."[46] Always before, that has been enough—that the world has meaning, and that it continues on when we depart it. Now we may actually have to choose—"live" without limits, or live *and die* with them.

When I wonder which I'd really choose, I try to think back on my life, imagine the moments I would relive in my mind were I on my deathbed. Almost without exception, those moments involve just the kind of rich meaning that the posthuman world would ask us to surrender: running a race with my lungs bursting, working with almost desperate dedication on a book, watching the birth of my daughter and knowing that she was her own mysterious self and my own lovely obligation. I think that I would not trade these things. I hope that I would not trade these things.

Is there some "goal" to our existence, some endpoint toward which we are heading? If there is, then perhaps it makes sense to speed up so we'll get there faster. What is it that we need all this extra intelligence to *figure out*? That we need all this new computer power to *do*? That robots will be capable of that we aren't? What is the task to which we must surrender so much? These questions sound preposterously large, ridiculously close to "What is the meaning of life?" They're not questions I'm comfortable asking; even many philosophers have given up on them, turning their profession into a branch of higher math. But when someone tells you, as the techno-utopians explicitly do, that they want to end your species as it has existed and replace it with something "higher" and "better," it seems useful to ask why.

It's not, I've tried to show, in order to cure the ill or feed the hungry. These things lie within our present powers or within the steady, foreseeable, noncontroversial progress of science and medicine. They don't require a posthuman future. So what does? What is our project? Why do we need to push the pace?

The men who propose this leap into the unbounded future don't seem to know themselves quite why they want to jump. Because "it will allow us a deeper understanding of what truly we are," says Rodney Brooks.[47] Because, says Gregory Stock, as our new biology allows us to "pierce the veneer of inside things, we may reach the naked soul of man."[48] Because, says J. Hughes, reengineered minds will "permit us to think more profound and intense thoughts."[49] These sound like things that people say to each other in the parking lot at a Phish concert, before they drop acid.

Perhaps I'm being unjust. Environmentalists, and I am one, have always been concerned with keeping the wonder of the present moment alive. With valuing what *is.* And perhaps, because of that bias, I've placed too much value on the world as we know it, failed to sufficiently appreciate the glories of the technoworld to come.

I've tried, I think. But the universal bliss of superenhanced semi-robots doesn't do much more for me than the harps and clouds of the church-school heaven. In truth, the longer you think about it, the sillier it becomes. Consider Mark Alan Walker, of the University of Toronto, who wrote an article for the *Transhumanist* on-line magazine with the imposing title "Prolegomena to Any Future Philosophy." He argues that philosophy has so far failed to "arrive at absolute knowledge, at a final theory of everything." In response, some of his colleagues have suggested that "the ambitions of philosophy ought to be scaled back to something more modest." Instead, he proposes that we use technology "to create beings who are as far removed from us in intelligence as we are from apes," and then wait for them to provide the answers. "It seems plausible to hypothesize that a creature who had a brain size of 2200 cubic centimeters ought to be more intelligent and have greater conceptual abilities than *Homo sapiens* with their measly 1300 cubic centimeters," he writes.

Perhaps they will then develop something called "hyperlanguages." They would, he writes, be "godlike." And then they could provide the theory of everything.[50]

In a similar vein, Lee Silver ends *Remaking Eden* by imagining some point in the far future, after the GenRich and the GenPoor have split apart, after their immortal descendants have gone into space, after their descendants have become as different from humans as "humans are from the primitive worms with tiny brains that first crawled along the earth's surface." Admittedly, he writes, "it is difficult to find words to describe the enhanced attributes of these special people. 'Intelligence' does not do justice to their cognitive abilities. 'Knowledge' does not explain the depth of their understanding. 'Power' is not strong enough to describe the control they have over technologies that can be used to shape the universe in which they live." So what do these sublime creatures do all the days of their endless lives? In his view, they dedicate their time to

> answering three deceptively simple questions that have been asked in every self-conscious generation of the past:
>
> "Where did the universe come from?"
>
> "Why is there something rather than nothing?"
>
> "What is the meaning of conscious existence?"[51]

Not to be impolite, but for this we trade our humanity? Sure, these questions are *important*, especially the last one. But they're not *all-important*. The techno-utopians ignore all the equally urgent queries, such as "What shall we have for dinner?" and "How are you feeling?" and "Can I give you a hand with that?" and "Do you think you could ever love me, too?"

Anyway, all those grander questions, especially the one concerning "the meaning of conscious existence," can only be usefully answered by people whose bodies eventually start to sag, by people who love and who grieve and who celebrate, by people who mourn and who know that they will someday die. There isn't a *right* answer to them, which we will find if only we summon enough brainpower.

There are only the sweet answers worked out over time by real humans in real life.

I went out for a run yesterday afternoon, just as I was finishing this book. As always, I was going pretty slowly, even when I sprinted—that's just what my body does. Dusk was settling on a late fall day, and so my fingers stung a little from the chill, my lungs ached a bit more than usual in the cold.

But it was one of those glorious evenings when the maple leaves were starting to cascade down at the slightest breeze, a shower of twisting red and orange. A full moon—a harvest moon—stood on the ridgeline. As the sunlight faded and lamps came on in living rooms and kitchens, I could see kids bent over homework, one more night's installment in the long process of building a solid mind. The blue glow of the television filtered out through lots of windows, but so did the smell of good food cooking, as it was cooking in a billion or two homes around the world at that same moment. A neighbor stacked wood on his pile, methodically, with much the same rhythm of bend and lift and twist that a billion or two other bodies had followed that day in rice paddy or workshop.

Doubtless there was pain and suffering and cruelty behind some of those walls, and just as surely there was joy and kindness, and sometimes in the same places. I ran by shabby trailer homes; some of my neighbors are too poor, and some, perhaps, too rich. To call the world enough is not to call it perfect or fair or complete or easy. But enough, just enough. And us in it.

NOTES

CHAPTER ONE: TOO MUCH

1. John M. Hoberman, *Mortal Engines: The Science of Performance and the Dehumanization of Sport* (New York: 1992), p. 72.
2. Ibid., pp. 136, 102; Sharon Begley, "Gold Medal Workouts," *Newsweek,* Dec. 17, 2001.
3. Mark Compton, "Enhancement Genetics: Let the Games Begin," *DNA Dispatch,* July 2001.
4. "More Giro Shocks Still to Come," *ProCycling,* March 5, 2002.
5. Amanda Swift, "The Sports Factor," ABC radio [Australia], July 12, 2001.
6. Ira Berkow, "This Lifter Is Fueled by Natural Power," *New York Times,* Feb. 6, 1994.
7. Rod Osher, "Hot Performances," Time.com, Sept. 6, 1999.
8. Michael Butcher, "Next: The Genetically Modified Athlete," *The Guardian* (England), Dec. 15, 1999.
9. Jere Longman, "Getting the Athletic Edge May Mean Altering Genes," *New York Times,* May 11, 2001.
10. Compton, "Enhancement Genetics."
11. "Erling Jevne: Down to Earth," *Skisport* magazine, translated and archived at *www.xcskiworld.com.*
12. "Jevne's Last Campaign," *www.langrenn.com,* Feb. 25, 2002.

13. J. D. Downing, "Golden Justice," *www.xcskiworld.com,* Feb. 25, 2002.
14. "Jevne's Last Campaign."
15. Carolyn Abraham, "Gene Pioneer Dreams of Human Perfection," *Toronto Globe and Mail,* Oct. 27, 2002.
16. Evelyne Shuster, "Of Cloned Embryos, Humans and Posthumans," *Yale Law Journal,* Sept. 13, 2001.
17. Richard Hayes, interview, *Wild Duck Review,* Summer 1999, p. 7.
18. Lee Silver, *Remaking Eden* (New York: 1999), p. 130.
19. Tim Beardsley, "A Clone in Sheep's Clothing," *Scientific American,* March 1997.
20. Rick Weiss, "Human Cloning's Numbers Game," *Washington Post,* Oct. 10, 2000.
21. Gregory Stock, *Redesigning Humans: Our Inevitable Genetic Future* (New York: 2002), p. 185.
22. "On Living Forever," interview with Michael West, *Ubiquity* magazine, *www.megafoundation.org,* June 2000.
23. "Cracking the Code," *Time,* July 3, 2000.
24. Francis S. Collins, Lowell Weiss, and Kathy Hudson, "Have No Fear, Genes Aren't Everything," *The New Republic,* June 25, 2001.
25. "Nature Beats Nurture," *The Times* (London), Feb. 13, 2001.
26. Tom Bethell, "Road Map to Nowhere," *American Spectator,* April 2001.
27. Richard Lewontin, "After the Genome, What Then?" *New York Review of Books,* July 19, 2001.
28. Keay Davidson, "Sticking a Pin in Genome Mappers' Balloon," *San Francisco Examiner,* July 5, 2000.
29. Francis Fukuyama, *Our Posthuman Future* (New York: 2002), p. 75.
30. Gina Kolata, "In Cloning, Failure Far Exceeds Success," *New York Times,* Dec. 11, 2001.
31. Stuart Newman, interview with author, Dec. 5, 2001.
32. David King, "David King on the Genome Announcement," *Genetic Crossroads Bulletin* 15 (Feb. 21, 2001).
33. Hayes, *Wild Duck Review.*
34. Newman interview.
35. "America's Next Ethical War," *The Economist,* April 14, 2001.
36. "Texas Researchers Clone Cat," BBC.news.com, Feb. 14, 2002.
37. "First Cloned Cat Is Born, Researchers Report," Reuters, Feb. 14, 2002.
38. Gina Kolata, "What's Warm and Fuzzy Forever? With Cloning, Kitty," *New York Times,* Feb. 15, 2002.
39. Brian Alexander, "You 2," *Wired,* Jan. 2001, p. 131.
40. Sylvia Pagan Westphal, "Chip Could Create Mass-Produced Clones," *www.newscientist.com,* Jan. 30, 2002.
41. "U.S. Firms Seek Mass Cloning of Chickens," Reuters, Aug. 15, 2001.
42. "Geneticists Create 'Enviropig,'" Reuters, June 24, 1999.
43. Gareth Cook, "Bunny Causes Outcry," *Los Angeles Times,* Sept. 19, 2000.

44. Gina Kolata, "Scientists Place Jellyfish Genes into Monkeys," *New York Times,* Dec. 23, 1999.
45. "Scientists Genetically Engineer a Monkey," Reuters, Jan. 11, 2001.
46. Daniel Kevles, *In the Name of Eugenics* (New York: 1985).
47. Silver, *Remaking Eden,* p. 245.
48. Sharon Begley, "Designer Babies," *Time,* Nov. 8, 1999.
49. Patrick Mooney, *The ETC Century* (Winnipeg: 2001), pp. 25–26.
50. Susan Watts, "Venter Reveals Clinton Apology over Human Genome," *BioMedNet News,* Feb. 13, 2002.
51. Rachel Nowak, "IVF Now Fastest Way to Get Pregnant," *New Scientist,* Jan. 16, 2002.
52. Alexander, "You 2," p. 131.
53. Ibid., p. 133.
54. "Human Embryo Created Through Cloning," CNN.com, Nov. 26, 2001.
55. Will Dunham, "World's First Genetically Altered Babies Born," Reuters, May 5, 2001.
56. Sheryl Gay Stolberg, "A Small Leap to Designer Babies," *New York Times,* Jan. 1, 2000.
57. Richard Hayes, "The Campaign to Ban Human Genetic Modification," April 2000.
58. Jonathan Knight, "Biology's Last Taboo," *Nature* 413 (Oct. 2001), pp. 12–15.
59. Lois Rogers, "Winston Patents Technique for Designer Sperm," *Sunday Times* (London), Dec. 10, 2000; Robin McKie, "Men Redundant? Now We Don't Need Women Either," *The Observer,* Feb. 10, 2002; Sheldon Krimsky, *Biotechnics and Society* (New York: 1991), p. 159.
60. Hayes, *Wild Duck Review.*
61. John Campbell and Gregory Stock, "A Vision for Practical Human Germline Engineering," in *Engineering the Human Germline,* ed. Campbell and Stock (Oxford, Eng.: 2000).
62. Knight, "Biology's Last Taboo."
63. John Campbell and Gregory Stock, "Engineering the Human Germline Symposium Summary Report," June 1998, *www.ess.ucla.edu/huge/report.*
64. Meredith Wadman, "Germline Gene Therapy Must Be Spared Excessive Regulation," *Nature* 392 (March 26, 1998), p. 317.
65. Robert Taylor, "Evolution Is Dead," *New Scientist,* Oct. 3, 1998, pp. 25–26.
66. Silver, *Remaking Eden,* pp. 1–3.
67. Fukuyama, *Posthuman Future,* pp. 80–81.
68. Brian Tokar, *Redesigning Life* (Boston: 2002), p. 147.
69. Frederic Golden, "Boy or Girl—Up to You," *Time,* Sept. 21, 1998.
70. Bill McKibben, *Hope, Human and Wild* (New York: 1995), p. 141.
71. Golden, "Boy or Girl."
72. Michael D. Lemonick, "Designer Babies," *Time,* Jan. 11, 1999.
73. Lee Silver, "A Quandary That Isn't," *Time,* Sept. 21, 1998.

74. Anthony G. Comuzzie and David B. Allison, "The Search for Human Obesity Genes," *Science* 280 (May 29, 1998), p. 1374.
75. Joseph Coates, "Treatment of Disease in the 21st Century—Towards the Manipulable Human," in Danish Council of Ethics, *Humans and Genetic Engineering in the New Millennium,* report of a conference held November 9, 1999, published by Danish Council of Ethics, 2000.
76. Lori Andrews, "Embryonic Confusion," *Genewatch,* Jan. 2001.
77. Jeremy Rifkin, *The Biotech Century* (New York: 1998), p. 151.
78. Daniel Q. Haney, "Designer Babies Just Genes Away," Associated Press, March 15, 2000.
79. Council on Genetics and Society, "The New Technology of Human Genetic Modification," May 2002, p. 16.
80. Silver, *Remaking Eden,* p. 238.
81. Martine Rothblatt, *Unzipped Genes* (Philadelphia: 1997), p. 49; Gregory S. Paul and Earl Cox, *Beyond Humanity: CyberEvolution and Future Minds* (Rockland, Mass.: 2001), p. 277; John Travis, "Eye-Opening News," *Science News* 151 (May 10, 1997), pp. 288–89.
82. Paul and Cox, *Beyond Humanity,* p. 277; Silver, *Remaking Eden,* p. 238.
83. Lauren Slater, "Dr. Daedalus: A Radical Plastic Surgeon Wants to Give You Wings," *Harper's,* July 2001.
84. Quoted in "IQ Linked with Amount of Gray Matter," BBC News.com, Sci/Tech, Nov. 5, 2001.
85. "The First Gene Marker for IQ?" *Science* 280 (May 1, 1998), p. 681.
86. Joe Tsien, "Building a Brainier Mouse," *Scientific American,* April 2000, pp. 65–69; Damien Broderick, *The Spike* (New York: 2001), p. 16.
87. Sharon Schmickle, "Intelligence Genes Prove Hard to Map," Scripps Howard News Service, March 10, 2002.
88. Fukuyama, *Posthuman Future,* p. 28.
89. Stock, *Redesigning Humans,* p. 101.
90. Bryan Appleyard, *Brave New Worlds* (New York: 1999) p. 88.
91. Philip Frazer, "Genetic Medicine Promises Baby Geniuses and Nightmares," *www.simulconference.com/public/swof/sowf/dispatches/dispatch 29.html.*
92. Ralph Brave, "Governing the Genome," *The Nation,* Dec. 10, 2001.
93. Rifkin, *Biotech Century,* p. 150.
94. Fukuyama, *Posthuman Future,* pp. 33–34.
95. Michael Holden, "Judges Urged to Consider Criminal Genes," Reuters, Oct. 2, 2000.
96. Nicholas Wade, "First Gene for Social Behavior Described in Whiskery Mice," *New York Times,* Sept. 9, 1997.
97. Stock, *Redesigning Humans,* p. 108.
98. Nicholas Wade, "Schizophrenia May Be Tied to Two Genes, Researchers Find," *New York Times,* July 4, 2002.
99. Dean Hamer, "The Heritability of Happiness," *Nature Genetics,* 14:125 (Oct. 1996).

100. Dean Hamer, "Tweaking the Genetics of Behavior," *Scientific American* Special Issue, "Your Bionic Future," Fall 1999, pp. 62–67.
101. Gregory Pence, *Who's Afraid of Human Cloning?* (New York: 1998), p. 168.
102. Alex Kuczynski, "In Quest for Wrinkle-Free Future, Frown Becomes Thing of the Past," *New York Times,* Feb. 7, 2002.
103. Rachel Zimmerman, "Botox Gives a Lift to These Soirees," *Wall Street Journal,* April 16, 2002.
104. Stock, *Redesigning Humans,* p. 9.
105. Barbara Carton, "For Modern Tots, Nothing Is Too Expensive," *Wall Street Journal,* Sept. 4, 2000.
106. Austin Bunn, "Terribly Smart," *New York Times Magazine,* March 24, 2002.
107. Rothblatt, *Unzipped Genes,* p. 6.
108. Rifkin, *Biotech Century,* p. 140.
109. Stephanie Mills, ed., *Turning Away from Technology* (San Francisco: 1997), p. 81.
110. Andrew Kimbrell, *The Human Body Shop* (San Francisco: 1993), p. 145.
111. Irene Sege, "A Good Egg," *Boston Globe,* March 3, 1999.
112. Carey Goldberg, "On Web, Models Auction Their Eggs to Bidders for Beautiful Children," *New York Times,* Oct. 23, 1999.
113. Lori B. Andrews, "Gen-Etiquette," *Christianity Today,* Oct. 1, 2001.
114. Tran T. Kim-Trang, letter to author, July 14, 2002.
115. Lester Thurow, *Creating Wealth: The New Rules* (New York: 1999), p. 33.
116. Stock, *Redesigning Humans,* p. 144.
117. Stock, "The Prospects for Human Germline Engineering," *Telepolis,* January 29, 1999.
118. Jon Cohen, "Designer Bugs," *The Atlantic,* July/Aug. 2002; "Scientists Fear Miracle of Biotech Could Also Breed a Monster," Agence France-Presse, Oct. 23, 2001.
119. Mooney, *ETC Century,* p. 42; Andrew Pollack, "Scientists Ponder Limits on Access to Germ Research," *New York Times,* Nov. 29, 2001.
120. Mark S. Frankel and Audrey R. Chapman, *Human Inheritable Genetic Modifications: Assessing Scientific, Ethical, Religious and Policy Issues* (Washington, D.C.: 2000) pp. 33–35.
121. Stock, *Redesigning Humans,* p. 132.
122. Jon Gordon, "Genetic Enhancement in Humans," *Science* 283 (1999), p. 2024.
123. Marc Lappé, "Ethical Issues in Manipulating the Human Germ Line," *Journal of Medicine and Philosophy* 16 (1991), pp. 621–39.
124. Kerry Cullinan, "Genetic Engineering Is the Future Racism," Online Health Service, *www.health-e.org.za,* April 9, 2001.
125. George Annas, speech at Beyond Cloning conference, Boston University, Sept. 21, 2001.

126. Silver, *Remaking Eden,* p. 11.
127. Ibid., pp. 4–6.
128. Ibid., p. 241.
129. Wesley Smith, "Cloning Reality," *National Review Online,* Jan. 31, 2001.
130. Freeman Dyson, "Progress in Religion," Edge 68, *www.edge.org/documents/archive/edge68.html.*
131. McKibben, *Hope, Human and Wild,* p. 130.
132. Fukuyama, *Posthuman Future,* p. 159.
133. Ibid., p. 7.
134. Ibid., p. 9.
135. Silver, *Remaking Eden,* pp. 9, 11.
136. Andrew Kimbrell, *The Human Body Shop* (New York: 1993), p. 74.
137. Jon Gordon, *Science,* March 25, 1999, p. 2023.
138. Gareth Cook, "Obesity in Mice Offers Proof of Cloning's Unpredictability," *Boston Globe,* March 1, 2002; Philip Cohen, "Cloned Animals Meet Early Deaths," *New Scientist,* Feb. 10, 2002.
139. Patricia Reaney, "Cloned Mice Show No Signs of Premature Aging," Reuters, Sept. 20, 2001.
140. Stuart Newman, "Genetic Enhancement—the State of the Technology," presentation at the Ecumenical Roundtable on Science, Technology, and the Church, Boston, April 23, 1998.
141. Stuart Newman, "The Role of Genetic Reductionism in Biocolonialism," *Peace Review* 12, no. 4 (2000), p. 523.
142. Leon Kass, "Moral Meaning of Genetic Technology," *Commentary,* Sept. 1999, p. 36.
143. Stuart Newman, personal interview, Dec. 5, 2001.
144. Fukuyama, *Posthuman Future,* pp. 79–80.
145. Rael, personal interview, June 7, 2002.
146. Daniel Koshland, "Ethics and Safety," in Campbell and Stock, *Engineering the Human Germline.*
147. Charles Krauthammer, "Crossing Lines," *The New Republic,* April 29, 2002.
148. Kass, "Moral Meaning," p. 35.
149. James Q. Wilson and Leon Kass, "The Ethics of Human Cloning," *www.theamericanenterprise.org/taema99q.*
150. Steve Connor, "Human Cloning Is Now Inevitable," *Independent* (London), Aug. 30, 2000.
151. David Abram, *The Spell of the Sensuous* (New York: 1996).
152. Frederick Engels and Karl Marx, *Communist Manifesto* (1848).
153. Albert Borgmann, *Crossing the Postmodern Divide* (Chicago: 1992), p. 99.
154. Charles Taylor, *Sources of the Self* (Cambridge, Mass.: 1989), p. 43.
155. Stock, *Redesigning Humans,* p. 120.

156. Sharon Begley, "In the New Field of Neurotheology, Scientists Seek the Biological Basis of Spirituality," *Newsweek,* May 7, 2001, p. 50.
157. Stock, *Redesigning Humans,* p. 120.
158. Ibid.
159. Hayes, *Wild Duck Review.*
160. Mihaly Csikszentmihalyi, *Beyond Boredom and Anxiety,* 25th anniversary edition (San Francisco: 2000).
161. Ibid., p. 8.
162. Ibid., p. 33.
163. Ibid., p. 39.
164. Ibid., p. 139.
165. Ibid., pp. 86–87.
166. Ibid., pp. 94–96.
167. Ibid., p. 194.
168. Ibid., p. 199.
169. David Strong and Eric Higgs, "Borgmann's Philosophy of Technology," in Eric Higgs, Andrew Light, and David Strong, eds. *Technology and the Good Life* (Chicago: 2000), pp. 27–28.
170. Albert Borgmann, *Technology and the Character of Contemporary Life* (Chicago: 1984), p. 207.
171. Silver, *Remaking Eden,* p. 237.
172. Ibid., p. 75.
173. Martha Nussbaum, "Brave Good World," *New Republic,* Dec. 4, 2000.
174. George Wright, "Scientists Create Pre-plucked Chicken," Reuters, May 21, 2002.
175. Silver, *Remaking Eden,* pp. 7–8.
176. "British Chief Rabbi Denounces Cloning," *Jerusalem Post,* Aug. 12, 2001.
177. Leon Kass, "Preventing a Brave New World," *New Republic,* May 17, 2001.
178. Bernd Heinrich, *Racing the Antelope* (New York: 2001), pp. 163–64.
179. Roger Angell, "The Interior Stadium," in Nicholas Dawidoff, ed., *Baseball: A Literary Anthology* (New York: 2002), p. 1.
180. Tom Verducci, "Steroids in Baseball," *Sports Illustrated,* June 3, 2002.
181. Allen Barra, "Steroids: The Cancer That's Growing Inside Baseball," *Salon.com,* May 31, 2002.
182. Rick Reilly, "The 'Roid to Ruin," *Sports Illustrated,* Aug. 21, 2000.
183. John McCardell, Middlebury College convocation, Sept. 9, 2001.
184. Timothy S. Sedone, "Born at the Destination: An Interview with Richard Rodriguez," *New England Review,* Summer 2001.
185. "On Living Forever," interview with Dr. Michael West, *Ubiquity* magazine, *www.megafoundation.org,* June 2000.

CHAPTER TWO: EVEN MORE

1. Jeffrey Sussman, "A Trillion Computers in a Drop of Water," *www.eurekalert.org,* Nov. 21, 2001.
2. John Markoff, "Hewlett Finds a Process to Make Chips Even Smaller," *New York Times,* Sept. 10, 2002.
3. Kenneth Chang, "Scientists Shrink Computing to Molecular Level," *New York Times,* Oct. 25, 2002.
4. Rodney Brooks, *Flesh and Machines* (New York: 2002), p. 27.
5. Charles Platt, "Superhumanism," *Wired,* Oct. 1995.
6. Ed Regis, *The Great Mambo Chicken and the Transhuman Condition* (New York: 1990) p. 154.
7. Karen Thomas, "The Future Is Already Written," *USA Today,* March 14, 2000.
8. Ray Kurzweil, *The Age of Spiritual Machines* (New York: 1999), p. 5.
9. Brooks, *Flesh and Machines,* p. 212.
10. Hans P. Moravec, *Robot: Mere Machine to Transcendent Mind* (Oxford, Eng.: 1999), p. 189.
11. Ibid., pp. 21–22.
12. Ibid., p. 25.
13. Hans Moravec, "Robots, Re-Evolving Mind," Speech at Carnegie Mellon University, Dec. 2000, accessed at *www.ri.cmu.edu/~hpm/project.archive/robot.papers/2000/Cerebrum/html.*
14. Kurzweil, *Spiritual Machines,* p. 4.
15. Damien Broderick, *The Spike* (New York: 2001), p. 35.
16. Moravec, *Robot,* p. 1.
17. Ray Kurzweil, "Accelerated Living," *PC Magazine,* Sept. 4, 2001.
18. Damien Cave, "Killjoy," Salon.com, April 10, 2000.
19. John Markoff, "The Increase in Chip Speed Is Accelerating, Not Slowing," *New York Times,* Feb. 4, 2002.
20. Jaron Lanier, "One Half of a Manifesto," Edge 74, *www.edge.org.*
21. Kurzweil, *Spiritual Machines,* p. 34.
22. Moravec, "Robots, Re-Evolving Mind."
23. Tom Logsdon, *The Robot Revolution* (New York: 1984), pp. 40–41.
24. Moravec, "Re-Evolving Mind."
25. Ibid.
26. Rodney Brooks, *Flesh and Machines: How Robots Will Change Us* (New York: 2000), p. 53.
27. Ibid., p. 64.
28. Anna Foerst, "Stories We Tell: Where Robotics and Theology Meet," Lecture at Columbia Center for the Study of Science and Religion, April 5, 2000.
29. Moravec, *Robot,* pp. 91–126.
30. Brooks, *Flesh and Machines,* p. 113.
31. "The Rise of Robots and the Decline of Humanity," *Earth Island Journal,* Summer 2002, p. 19.

32. Patrick Mooney, *The ETC Century* (Winnipeg: 2001), p. 58.
33. Anne Eisenberg, "Designers Take Robots out of Human Hands," *New York Times,* Feb. 28, 2002.
34. "Inventions of the Year, 2001," *Time,* Nov. 9, 2001.
35. John Farrell, "The Monster from . . . Waltham," *National Review Online,* Sept. 1, 2000.
36. Brooks, *Flesh and Machines,* p. 200.
37. Moravec, *Robot,* p. 73.
38. Kurzweil, *Spiritual Machines,* p. viii.
39. Ibid., p. 105.
40. Broderick, *Spike,* p. 18.
41. Charles Krauthammer, "Deep Blue Funk," *Time,* Feb. 26, 1996.
42. Ashley Dunn, "If Deep Blue Wrote *Hamlet,* Would It Change the Endgame?" *New York Times,* May 14, 1997.
43. Regis, *Mambo Chicken,* pp. 15–16.
44. Mooney, *ETC Century,* p. 44.
45. B. C. Crandall, *Nanotechnology: Molecular Speculations on Global Abundance* (Cambridge, Mass.: 1996) p. 23.
46. Ibid.; Regis, *Mambo Chicken,* p. 140.
47. Crandall, *Nanotechnology: Molecular Speculations,* p. 25.
48. "Nano Technology," *Business Week Online,* March 26, 2002.
49. Ibid.
50. Regis, *Mambo Chicken,* pp. 120–21.
51. "Nanotechnology," *Business Week Online,* March 26, 2002.
52. Rodney Brooks, replying to Jaron Lanier, Edge 74, *www.edge.org.*
53. Nicholas Thompson, "Downsizing," *Washington Monthly,* Oct. 2000.
54. R. Michael Perry, *Forever for All* (2000), p. 262.
55. Mooney, *ETC Century,* p. 46.
56. "The Nanotech Economy," *Nanotechnology* magazine overview section, *www.nanozine.com.*
57. Harry Chesley, "Early Applications," in B. C. Crandall, ed., *Nanotechnology: Research and Perspectives* (Cambridge, Mass.: 1992).
58. Richard Crawford, "Cosmetic Nanosurgery," in Crandall, ed., *Nanotechnology: Research and Perspectives,* p. 72.
59. Ibid., p. 63–64.
60. Edward M. Reifman, "Diamond Teeth," in Crandall, *Nanotechnology: Molecular Speculations,* p. 84.
61. "The Incredible Shrinking World of Eric Drexler," *Red Herring,* Aug. 1995.
62. Barnaby J. Feder, "Tiny Technologies Slip Unseen into Daily Life," *New York Times,* March 11, 2002.
63. "Nanotechnology," *Business Week Online,* March 26, 2002.
64. Feder, "Tiny Technologies."
65. "Nanotechnology," March 26, 2002.
66. Tiffany Kary, "Nanotech More Science Than Fiction," *ZDNetNews,* Feb. 11, 2002.

67. Broderick, *Spike,* p. 67.
68. Nicholas Thompson, "Downsizing," *Washington Monthly,* Oct. 2000.
69. Speech of the President, Caltech, Jan. 21, 2000, *www.whitehouse.gov.*
70. *www.foresight.org/updates.update40.1.html,* Dec. 13, 2000.
71. "Bill Clinton Calls Many Political Leaders out of Touch with the Acceleration of Technology at Fortune Summit," *KurzweilAI.net,* Aug. 3, 2001.
72. Declan McCullagh, "Big Money for Tiny Tech," *ZDNetNews,* Sept. 16, 2002.
73. *www.smalltimes.com/document_display.* Document_id=2736.
74. "The Nanotechnology Economy," Nanozine.com; Mooney, *ETC Century,* p. 51.
75. Amara D. Angelica, "Texas Thinks Small, Plans Nanotech Corridor," *KurzweilAI.net,* June 11, 2001.
76. Broderick, *Spike,* p. 151.
77. Ibid., p. 153.
78. "When Will Nanotechnology Arrive?" Nanozine.com
79. Mooney, *ETC Century,* p. 51.
80. Kary, "Nanotech."
81. Mooney, *ETC Century,* p. 53.
82. Broderick, *Spike,* p. 11.
83. Kenneth Chang, "Can Robots Rule the World? Not Yet," *New York Times,* Sept. 12, 2000.
84. Ray Kurzweil, "Live Forever," *Psychology Today,* Feb. 2, 2001.
85. Steve Alan Edwards, "Mind Children: Extropians," *21C: Scanning the Future,* Spring 1997.
86. Broderick, *Spike,* p. 203.
87. Kenneth Chang, "Using Robotics, Researchers Give Upgrade to Lowly Rats," *New York Times,* May 2, 2002.
88. Mooney, *ETC Century,* p. 58.
89. Ian Pearson, "The Future of Human Evolution," *Sphere—BT's Innovation and Technology Ezine,* March 15, 2000.
90. Bill Joy, "Why the Future Doesn't Need Us," *Wired,* April 2000.
91. George Gilder and Richard Vigilante, "Stop Everything . . . It's Techno Horror," *American Spectator,* March 2001.
92. Joy, "Why the Future Doesn't Need Us."
93. Ibid.
94. Damien Cave, "Kill Joy," Salon.com, April 10, 2000.
95. Joy, "Why the Future Doesn't Need Us."
96. Robert Freitas, "Some Limits to Global Ecophagy by Biovorous Nanoreplicators, with Public Policy Recommendations," *www.Zyvex.com,* April 2000.
97. Alex Roslin, "Cyber-Doomsayers Offer Chilling Vision," *Montreal Gazette,* Oct. 14, 2001.
98. Broderick, *Spike,* pp. 112–13.
99. Freitas, "Some Limits."

100. Joy, "Why the Future Doesn't Need Us."
101. "Scientists Create Live Polio Virus," *New York Times,* July 12, 2002.
102. Gregory S. Paul and Earl Cox, *Beyond Humanity: CyberEvolution and Future Minds* (Rockland, Mass.: 2001), p. 248.
103. Moravec, *Robot,* p. 132.
104. Pearson, "Future of Human Evolution."
105. Moravec, *Robot,* pp. 139–40.
106. Erazim Kohák, *The Embers and the Stars* (Chicago: 1984), p. 212.
107. Wendell Berry, "The Boundary," in Berry, *The Wild Birds: Six Stories of the Port William Membership* (New York: 1986), p. 88.
108. Wendell Berry, *What Are People For?* (Washington: 1996), pp. 190–91.
109. Jim Wilson, "Shrinking Micromachines," *Popular Mechanics,* Nov. 1997.
110. "Nanotechnological Pursuit of Happiness," Nanozine.com.
111. Paul and Cox, *Beyond Humanity,* p. 98.
112. Marvin Minsky, "Our Roboticized Future," in Marvin Minsky, ed., *Robotics* (New York: 1985), p. 304.
113. Wendell Berry, "Discipline and Hope," in Wendell Berry, *Recollected Essays* (San Francisco: 1981), pp. 179–80.
114. Kurzweil, *Spiritual Machines,* p. 97.
115. Robert Freitas, "Birth of the Cyborg," in Minsky, ed., *Robotics.*
116. David F. Gallagher, "Robots Find a Muse Other Than Mayhem," *New York Times,* May 30, 2002.
117. Dan Milmo, "Mentorn's Robots Dominate the Globe," *Guardian* (U.K.), Jan. 11, 2002.
118. Tom McKendree, "Nanotech Hobbies," in B. C. Crandall ed., *Nanotechnology: Molecular Speculations,* p. 142.
119. Freitas, "Birth of the Cyborg," p. 164.
120. N. Katherine Hayles, *How We Became Posthuman* (Chicago: 1999), p. 3; Kevin Kelly, quoted in James Bell, "Technotopia and the Death of Nature," *Earth Island Journal,* Summer 2002.
121. Paul and Cox, *Beyond Humanity,* p. 345.
122. David Abram, *Spell of the Sensuous* (New York: 1996), p. 23.
123. Brooks, *Flesh and Machines,* p. 236.
124. Moravec, "Re-Evolving Mind."
125. Stock, *Redesigning Humans,* p. 4.
126. Robin McKie, "I'm the Human Genome, Says 'Darth Venter' of Genetics," *Observer* (England), April 28, 2002.
127. Brooks, *Flesh and Machines,* p. 64.
128. Paul and Cox, *Beyond Humanity,* pp. 314–15.
129. Vernor Vinge, "Technological Singularity," *www.singularity.org,* 1993.
130. "Nanotechnological Pursuit of Happiness," Nanozine.com.
131. J. Storrs Hall, "What I Want to Be When I Grow Up Is a Cloud," *KurzweilAI.net,* July 6, 2001.
132. David Pearce, *The Hedonistic Imperative, www.hedweb.com.*
133. Warren Wengar, "Of Men and Machines," *Transhumanity* magazine, March 15, 2002.

134. Frank Tipler, *The Physics of Immortality* (New York: 1994), p. xi.
135. Broderick, *Spike,* p. 128.
136. Wendy Bousfield, "From the Island of Dr. Moreau to Lives of the Monster Dogs," *web.syr.edu/~blbousfi/UPLIFTEX22,* May 30, 1999.
137. Greg Cox, *Star Trek: The Eugenics Wars* (New York: 2001), p. 2.
138. Tom Logsdon, *The Robot Revolution* (New York: 1990), p. 45.
139. Stock, *Redesigning Humans,* pp. 19–20.
140. Patricia Leigh Brown, "Blinded by Science," *New York Times,* July 14, 2002.
141. Arthur C. Clarke, *The City and the Stars* (London: 1956), p. 106.

CHAPTER THREE: ENOUGH?

1. Ed Regis, *The Great Mambo Chicken and the Transhuman Condition* (New York: 1990), p. 167.
2. Robert Ettinger, *The Prospect of Immortality* (Garden City, N.Y.: 1964), p. 148.
3. Regis, *Mambo Chicken,* p. 160.
4. John Brockman, introduction to "Consciousness Is a Big Suitcase: A Talk with Marvin Minsky," Edge 3rd Culture, *www.edge.org.*
5. Marvin Minsky, "Will Robots Inherit the Earth?" *Scientific American,* Oct. 1994.
6. Gregory S. Paul and Earl Cox, *Beyond Humanity: CyberEvolution and Future Minds* (Rockland, Mass.: 2001), p. 281.
7. Regis, *Mambo Chicken,* p. 146.
8. George Gilder and Richard Vigilante, "Stop Everything . . . It's Techno Horror," *American Spectator,* March 2001.
9. "The Morality of Extremism," Extropian Web digest, April 18 and 19, 2002. It should be noted that the discussion of violence was a rare exception on this particular Web forum, which is generally starry-eyed but peaceful.
10. Michel Houellebecq, *The Elementary Particles* (New York: 2000), p. 262.
11. *Nanotechnology* magazine, overview, *www.nanozine.com.*
12. Damien Broderick, *The Spike* (New York: 2001), p. 83.
13. Heidi B. Perlman, "Technique Clones Endangered Animals," *San Francisco Examiner,* Oct. 9, 2000.
14. Paul and Cox, *Beyond Humanity,* pp. 298–99.
15. Ibid., p. 429.
16. Robert Nozick, *The Examined Life* (New York: 1990), pp. 236–38.
17. Lester Brown, *Eco-Economy: Building an Economy for the Earth* (New York: 2001), pp. 104–5.
18. Ibid., p. 118.
19. Erica Klarreich, "Population Set to Decline," *Nature,* Aug. 2, 2001.
20. Lee Silver, *Frontline* interview, 1999 at *www.pbs.org/wgbh/pages/frontline/shows/fertility.*

21. Ray Kurzweil, "Accelerated Living," *PC Magazine,* Sept. 4, 2001.
22. Hans P. Moravec, *Robot: Mere Machine to Transcendent Mind* (Oxford, Eng.: 1999), p. 9.
23. Errol Morris, *Fast, Cheap & out of Control,* Sony pictures, released Oct. 1997.
24. H. Keith Henson, "Trivial (Uses of) Nanotechnology," in B. C. Crandall and James Lewis, ed., *Nanotechnology: Research and Perspectives* (Cambridge, Mass.: 1992), p. 128.
25. Rodney Brooks, *Flesh and Machines* (New York: 2002), p. 229.
26. Ibid., p. 125.
27. Albert Borgmann, *Technology and the Character of Contemporary Life* (Chicago: 1987), p. 140.
28. Nicholas Rescher, *Unpopular Essays on Technological Progress* (Pittsburgh: 1980), p. 11.
29. Gregory Stock, *Redesigning Humans: Our Inevitable Genetic Future* (New York: 2002), p. 150.
30. Leroy Walters and Julie Gage Palmer, *Ethics of Human Gene Therapy* (Oxford, Eng.: 1997), p. 11.
31. Burke K. Zimmerman, "Human Germ Line Intervention," *Journal of Medicine and Philosophy* 16 (1991), pp. 593–612.
32. John Campbell and Gregory Stock, "Engineering the Human Germline Symposium Summary Report," June 1998, *www.ess.ucla.edu/huge/report.edu,* p. 13.
33. Richard Hayes, interview, *Wild Duck Review,* Summer 1999.
34. Barbara Katz Rothman, *Genetic Maps and Human Imaginations: The Limits of Science in Understanding Who We Are* (New York: 1998), p. 175.
35. Stuart Newman, interview with author, Dec. 5, 2001.
36. Sheryl Gay Stolberg, "Despite Ferment, Gene Therapy Progresses," *New York Times,* June 6, 2000.
37. Gina Kolata, "Scientists Report First Success of Gene Therapy," *New York Times,* April 28, 2000.
38. Sheryl Gay Stolberg, "Despite Ferment, Gene Therapy Progresses."
39. Andrew Pollack, "F.D.A. Halts 27 Gene Therapy Trials After Illness," *New York Times,* Jan. 15, 2003.
40. Richard Hayes, speech, Kennedy School of Government, Harvard University, May 9, 2000.
41. David King, "Technologies Which May Contribute to Human Genetic Engineering and Eugenics: A Brief Summary," handout prepared for workshop on the social impacts of the new human genetics technologies, UC Berkeley, Nov. 21, 1998, p. 1.
42. Zimmerman, "Human Germ Line Intervention," p. 598.
43. Lee Silver, "Making Babies," *Frontline,* 1999 (transcript on Web site: Frontline.org).
44. Sheryl Gay Stolberg, "A Genetic Future Is Both Tantalizing and Disturbing," *New York Times,* Jan. 1, 2000.
45. Hayes, *Wild Duck Review.*

46. Lee Silver, *Remaking Eden,* (New York: 1999), p. 229.
47. Ibid., p. 242.
48. Kyla Dunn, "Cloning Trevor," *Atlantic Monthly,* June 2002.
49. Robert Weinberg, "Of Clones and Clowns," *Atlantic Monthly,* June 2002, p. 57.
50. George W. Bush, Remarks on Human Cloning Legislation, April 10, 2002, *www.whitehouse.gov.*
51. Sheryl Gay Stolberg, "House Republicans Press Senate on Cloning," *New York Times,* May 16, 2002.
52. "American's Next Ethical War," *The Economist,* April 14, 2001.
53. Richard Hayes, "Designer Children," *Christian Social Action,* Oct. 1999, p. 16.
54. Leon Kass, "Preventing a Brave New World," *The New Republic,* May 17, 2001.
55. Richard Hayes, "Missed Message: Deeper Analysis of the President's Council on Bioethics Report," *Genetic Crossroads* 22 (July 11, 2002).
56. Weinberg, "Of Clones and Clowns," p. 59.
57. Leon Kass, "Preventing a Brave New World," *The New Republic,* May 17, 2001.
58. Wesley J. Smith, "Thank God for Cloning," *National Review Online,* Aug. 3, 2001.
59. "Using a Patient's Own Bone Marrow Cells Can Help an Ailing Heart," *www.eurekalert.com,* Nov. 13, 2001.
60. Warren King, "High on the Future: Already Saving Lives, Stem-Cell Research May Soon Be in Full Swing," *Seattle Times,* Aug. 20, 2001.
61. Tom Arnold, "Canadian Researchers Have Been Able to Rebuild Nerves in Rats," *National Post* (Ottawa), Aug. 17, 2001.
62. Leon Kass, "The Moral Meaning of Genetic Technology," *Commentary,* Sept. 1999.
63. Andrew Pollack, "Drugs to Spur New Cells, and Without the Politics," *New York Times,* Dec. 13, 2001.
64. Lawrence K. Altman, "Cancer Doctors See New Era of Optimism," *New York Times,* May 22, 2001.
65. Denis Grady, "A Final Option—an Experimental Alternative to Chemotherapy," *New York Times,* April 9, 2002; Laurie Tarkan, "Treatments in the Wings: New Drugs Could Replace Tamoxifen," *New York Times,* April 9, 2002.
66. Natalie Angier, "Getting Used to Long Life with Cancer," *New York Times,* April 9, 2002.
67. Robert Taylor, "Evolution Is Dead," *New Scientist,* Oct. 3, 1998.
68. Interview, Stuart Newman, Sept. 17, 2002.
69. Andrew Kimbrell, *The Human Body Shop* (San Francisco: 1993) p. 119.
70. Michael Conlon, "Genetically Selected Baby Born Free of Alzheimer's," *Reuters,* Feb. 26, 2002.
71. Margaret Talbot, "Jack or Jill," *Atlantic Monthly,* March 2002, p. 25.

72. Silver, *Remaking Eden,* pp. 200–203.
73. Stock, *Redesigning Humans,* p. 177.
74. Michael Hanlon, "Why You Won't Need to Have Sex to Make a Baby," *Montreal Express,* Oct. 25, 2000.
75. Francis Fukuyama, "How to Regulate Science," *Public Interest,* Winter 2002.
76. Royal Canadian Commission on Reproductive Technology, p. 942.
77. J. Madeline Nash, "Grains of Hope," *Time,* July 31, 2000.
78. Michael Pollan, "The Great Yellow Hype," *New York Times Magazine,* March 4, 2001.
79. Ronald Bailey, "Where's the Golden Rice?" *Reason,* Dec. 6, 2000.
80. Bill Joy, "Why the Future Doesn't Need Us," *Wired,* April 2000.
81. Carmelo Ruiz-Marrero, "Biotech's Third Generation," *Corpwatch,* April 5, 2002.
82. Pollan, "Great Yellow Hype."
83. Davinder Sharma, "Biotech—Exploiting the Poor and Hungry," speech delivered July 2, 2001, at the National Convention on Biotech and Hunger, Hyderabad, India.
84. Mark Winston, *Travels in the Genetically Modified Zone* (Cambridge, Mass.: 2002), pp. 225–26.
85. Jules Pretty, "The Magic Bean," *Resurgence* 210 (Jan./Feb. 2002), p. 21.
86. Craig Holdredge and Steve Talbot, "Sowing Technology," *Sierra,* July/Aug. 2001, p. 34.
87. "Against the Grain," interview with Tewolde Egziabher, *Sierra,* July/Aug. 2001, p. 30.
88. Ernest Becker, *The Denial of Death* (New York: 1973), pp. 34–35.
89. Interview with Rael and other members of his group, June 7, 2002.
90. Stephen Hall, "The Recycled Generation," *New York Times Magazine,* Jan. 30, 2000.
91. "On Living Forever," interview with Dr. Michael West, *Ubiquity* magazine, *www.megafoundation.org,* June 2000.
92. Aaron Zitner, "Clones, Free Love, and UFOs," *Los Angeles Times,* March 5, 2002.
93. Joannie Fischer, "Scientists Have Finally Cloned a Human Embryo," *U.S. News & World Report,* Dec. 3, 2001.
94. Raja Misha, "Scientific Maverick on a Quest for Old Age 'Cure,'" *Boston Globe,* Aug. 19, 2001.
95. "On Living Forever," *Ubiquity.*
96. Hall, "Recycled Generation."
97. Stock, *Redesigning Humans,* p. 78.
98. Damien Broderick, Extropian Web digest, April 3, 2002.
99. Interview with Robert Freitas, Jr., Foresight Update 38, p. 2, *www.foresight.org.*
100. Becker, *Denial of Death,* pp. 34–35, 96.

101. Paul and Cox, *Beyond Humanity,* pp. xiii, 272.
102. Leonard Hayflick, *How and Why We Age* (New York: 1994), p. 57.
103. Pat Mooney, *The ETC Century* (Winnipeg: 2001), p. 67.
104. Hayflick, *How and Why,* p. 84.
105. Liz F. Kay, "We Can Live Far Longer, Study Says," *Los Angeles Times,* May 10, 2002.
106. Hayflick, *How and Why,* pp. 286–87.
107. Ibid., p. 84.
108. Carol Kahn, *Beyond the Helix* (New York: 1985), p. 16.
109. Ibid., p. 17.
110. Hayflick, *How and Why,* pp. 117–20.
111. Ibid., pp. 134–36.
112. Paul and Cox, *Beyond Humanity,* pp. 290–91.
113. Hayflick, *How and Why,* p. 94.
114. James S. Goodwin and Jean M. Goodwin, "Impossibility in Medicine," in Philip J. Davis and David Park, eds., *No Way: The Nature of the Impossible* (New York: 1987), p. 48.
115. Hayflick, *How and Why,* p. 100.
116. Richard Saltus, "The Secret of Life," *Boston Globe,* May 22, 2001.
117. Hayflick, *How and Why,* p. 84.
118. Ibid., pp. 273–74.
119. Kahn, *Beyond the Helix,* pp. 34, 68.
120. Robin Mckie, "Discovery of Methuselah Gene Unlocks Secret of Long Life," *Observer* (England), Feb. 3, 2002.
121. Patricia Reaney, "Genetic Tinkering Makes Roundworms Live Longer," Reuters, March 7, 2001.
122. Damien Broderick, *The Last Mortal Generation* (Sydney: 1999), p. 4.
123. Laura Johannes, "The Surprising Rise of a Radical Diet," *Wall Street Journal,* June 3, 2002.
124. Kahn, *Beyond the Helix,* p. 99.
125. Johannes, "Surprising Rise."
126. Geoffrey Cowley, "The Biology of Aging," *Newsweek* special edition, Fall/Winter 2001, p. 12.
127. Gina Kolata, "Pushing Limits of Human Life Span," *New York Times,* March 9, 1999.
128. Kyla Dunn, "Cloning Trevor," *Atlantic Monthly,* June 2002, p. 46.
129. "Nanotechnology Health," Nanozine.com.
130. Bruce Schechter, "They've Seen the Future and Intend to Live It," *New York Times,* July 16, 2002.
131. Regis, *Mambo Chicken,* p. 85.
132. Ettinger, *Prospect of Immortality,* p. 85.
133. Regis, *Mambo Chicken,* p. 86.
134. Ettinger, *Prospect of Immortality,* p. 127.
135. Regis, *Mambo Chicken,* p. 5.
136. Alcor Life Extension Foundation, *www.Alcor.org.*

137. R. Michael Perry, *Forever for All* (Parkland, Fla.: 2000), p. 43; James Halperin, *The First Immortal* (New York: 1998), p. 161.
138. Halperin, *First Immortal,* pp. 277, 289.
139. Ettinger, *Prospect of Immortality,* p. 184.
140. Stock, *Redesigning Humans,* p. 96.
141. Broderick, *Last Mortal Generation,* p. 202.
142. Interview with Robert Ettinger, *Transhumanity* magazine, March 22, 2002, *www.transhumanism.com/archive.*
143. Broderick, *Last Mortal Generation,* p. 201.
144. Ibid., p. 57.
145. Leon Kass, "L,Chaim and Its Limits: Why Not Immortality," in Philip Zaleski, ed., *Best Spiritual Writing of 2002* (San Francisco: 2002), p. 98.
146. Perry, *Forever for All,* p. 440.
147. Ettinger, *Prospect of Immortality,* p. 174.
148. Broderick, *Last Mortal Generation,* p. 68.
149. Becker, *Denial of Death,* p. 26.
150. Robert Pack, *Affirming Limits: Essays on Mortality, Choice, and Poetic Form* (Amherst, Mass.: 1985) p. 107.
151. Martin Luther King, Jr., speech, Detroit, June 23, 1963, *www.african americans.com/mlkjrspeechmenu.htm.*
152. Arthur C. Clarke, *The City and the Stars* (London: 1976), p. 92.
153. "On Living Forever," *Ubiquity* magazine, *www.megafoundation.org,* June 2000.

CHAPTER FOUR: IS ENOUGH POSSIBLE?

1. Rodney Brooks, *Flesh and Machines* (New York: 2002), p. 238.
2. Gregory Stock, *Metaman: The Merging of Humans and Machines into a Global Superorganism* (New York: 1993), p. 168.
3. Lee M. Silver, *Remaking Eden* (New York: 1999), pp. 240–41.
4. N. Katherin Hayles, "The Life Cycle of Cyborgs," in Chris Hables Gray, ed., *The Cyborg Handbook* (New York: 1995), p. 327.
5. Sandy Stone, "Split Subjects, Not Atoms," in Hables Gray, ed., *Cyborg Handbook,* p. 393.
6. Gregory Stock, *Redesigning Humans* (Boston: 2002), p. 151.
7. Michael Lorrey, press release, Center for Transhuman Development, April 30, 2002, *www.cryonet.org/cgi-bia/dsp-cgimsg=19003.*
8. Silver, *Remaking Eden,* p. 8.
9. Lois Rogers and Joe Lauria, "Lawyer in Fresh Fight to Clone Dead Baby Son," *Sunday Times* (London), Aug. 19, 2001.
10. David Akin, "With Stem Cell Researchers Roaming the Earth in Search of Liberal Research Laws, Governments Face a Regulatory Nightmare," *Toronto Globe and Mail* , Feb. 6, 2002.
11. Robert Weinberg, "Of Clones and Clowns," *Atlantic Monthly,* June 2002.

12. Paul Recer, "World 'Wide Open' for Cloning," Associated Press, Dec. 14, 2001.
13. Stock, *Redesigning Humans,* p. 160.
14. J. Hughes, "Relinquishment or Regulation," *www.changesurfer.com/acad/relreg/pdF.*
15. Ralph Merkle interview, *Ubiquity* magazine, July 11–17, 2000. *www.acm.org/ubiquity/interviews/r-merkle-html.*
16. George Gilder and Richard Vigilante, "Stop Everything . . . It's Techno Horror," *American Spectator,* March 2001.
17. Gregory S. Paul and Earl Cox, *Beyond Humanity: CyberEvolution and Future Minds* (Rockland, Mass.: 2001), p. 435.
18. Antonio Regalado and Meera Louis, "Ethical Concerns Block Patents of Useful Embryonic Advances," *Wall Street Journal,* Aug. 20, 2001.
19. Tom Abate, "Tech Giants Jumping into Biotech," *San Francisco Chronicle,* Oct. 9, 2000.
20. Kurzweil, speech at National Cathedral, Washington, D.C., Nov. 19, 2001, *www.KurzweilAI.net.*
21. Silver, *Remaking Eden,* p. 11.
22. Gilder and Vigilante, "Stop Everything."
23. Jon Gordon, "Genetic Enhancement in Humans," *Science* 283 (March 26, 1999), p. 2024.
24. Ralph Brave, "Governing the Genome," *The Nation,* Dec. 10, 2001.
25. Hans P. Moravec, *Robot: Mere Machine to Transcendent Mind* (Oxford, Eng.: 1999), p. 7.
26. Stock, *Redesigning Humans,* p. 39.
27. Marc Olshan, "Modernity and Folk Society, in Donald B. Kraybill and Marc Olshan, eds., *The Amish Struggle with Modernity* (Hanover, N.H.: 1994).
28. Diane Zimmerman Umble, "Amish on the Line: The Telephone Debates," in Kraybill and Olshan, eds., *Amish Struggle,* pp. 105–6.
29. John Hostetler, *Amish Society* (Baltimore: 1993), pp. 96–97.
30. Gene Logsdon, *At Nature's Pace* (New York: 1994), pp. 142–43.
31. Samuel M. Wilson, "The Emperor's Giraffe," *Natural History,* Dec. 1992.
32. John K. Fairbank, Edwin O. Reischauer, and Albert M. Craig, *East Asia: Tradition and Transformation* (Cambridge, Mass.: 1973), p. 183.
33. Shih-Shan Henry Tsai, *The Eunuchs in the Ming Dynasty* (Albany, N.Y.: 1996), p. 156.
34. Louise Levathes, *When China Ruled the Seas* (New York: 1994), p. 100.
35. Nicholas D. Kristof, "1492: The Prequel," *New York Times,* Jan. 1, 2000.
36. Ibid.
37. Peter C. Perdue, "Technology in World History," paper presented at New England Regional World History Association, *whc.neu.edu/ner_what/technology/essay3html;* Tsai, *Eunuchs,* p. 164.
38. Kristof, "1492."
39. Fairbank, Reischauer, and Craig, *East Asia,* p. 178.

40. Noel Perrin, *Giving Up the Gun* (Boston: 1979), pp. 9–10; Arnold Pacey, *Technology in World Civilization* (Cambridge, Mass.: 1991), p. 89.
41. Perrin, *Giving Up the Gun,* p. 11.
42. Ibid., pp. 24–25.
43. Ibid., pp. 70–71.
44. Ibid., pp. 24–25.
45. Ibid., p. 12.
46. Ibid., p. 78.
47. Fairbank, Reischauer, and Craig, *East Asia,* p. 410.
48. Perrin, *Giving Up the Gun,* p. 92.
49. FIS Official Says Doping May Be on the Way Down," *www.xcskiworld.com,* Nov. 12, 2002.
50. "Federal Grand Jury Probes Human Cloning Claims," WHAW Radio News, *WHAWRadio.com/RadioNews18.htm,* August 12, 2001, quoting *Charleston Daily Mail.*
51. Michele Kambas, "Cypriot Researcher Sees Human Clone in Four Months," *Reuters,* Oct. 5, 2001.
52. Sheryl Gay Stolberg, "House Republicans Press Senate on Cloning," *New York Times,* May 16, 2002.
53. "Italy Lawmakers Against Human Clones," Associated Press, March 10, 2001; "First Cloned Baby Is 'Son of a Rich Arab,'" *www.ananova.com,* April 9, 2002.
54. Roger Highfield, "Cloned Baby Row Doctor Has Run out of Patients," *Daily Telegraph* (London), April 27, 2002.
55. "Woman to Bear a Clone, a Doctor Says," *New York Times,* Nov. 27, 2002.
56. Robin Marantz Henig, "Adapting to Our Own Engineering," *New York Times,* Dec. 16, 2002.
57. Maxwell Mehlman, "Regulating Genetic Enhancement," *Wake Forest Law Review* 34 (Fall 1999), p. 714.
58. Stock, *Redesigning Humans,* p. 60.
59. "Harry and Louise Amok," *American Prospect,* June 3, 2002.
60. Jackie Stevens, "P.R. for the Book of Life," *www.nation.com,* Nov. 26, 2001.
61. Wesley J. Smith, "Science or Propaganda?" *Weekly Standard,* Jan. 24, 2002.
62. James Watson, in Campbell and Stock, *Engineering the Human Germline* (New York: 2000).
63. Marcy Darnovsky, "The Case Against Designer Babies," in Brian Tokar, *Redesigning Life* (Boston: 2002), p. 138.
64. Meredith Wadman, "Germline Gene Therapy Must Be Spared Excessive Regulation," *Nature* 342 (March 26, 1998), p. 317.
65. Marcy Darnovsky, "Koppel Does Cloning" (San Francisco, 1999), available from Center for Genetics and Society, *www.genetics-and-society.org.*

66. Stephen Hall, "The Recycled Generation," *New York Times Magazine,* Jan. 30, 2000.
67. "On Living Forever," interview with Michael West, *Ubiquity* magazine, *www.megafoundation.org,* June 2000.
68. Tom Pelton, "Rivals Embody Opposite Sides of Cloning Debate," *Baltimore Sun,* April 7, 2002.
69. John Maddox, "The Case for the Human Genome," *Nature* 352 (July 4, 1991), p. 12.
70. Marvin Minsky, "Our Roboticized Future," in Marvin Minsky, ed., *Robotics* (New York: 1985), p. 292.
71. Sheryl Gay Stolberg, "Study Says Clinical Guides Often Hide Ties of Doctors," *New York Times,* Feb. 6, 2002.
72. "Just How Tainted Has Medicine Become?" *Lancet* 359 (April 6, 2002), p. 9313.
73. Deborah Nelson and Rick Weiss, "How We Uncovered the Hidden Fatality in a Clinical Trial," *Newsletter of the National Association of Science Writers,* Spring 2000.
74. Jonathan Embank, Correspondence, *Nature* 392 (April 16, 1998), p. 645.
75. Tom Abate, "Nobel Winner's Theories Raise Uproar in Berkeley," *San Francisco Chronicle,* Nov. 13, 2000.
76. Barbara Ehrenreich, "Double Helix, Single Guy," *New York Times Book Review,* Feb. 24, 2002.
77. Carolyn Abraham, "Gene Pioneer Dreams of Human Perfection," *Toronto Globe and Mail,* Oct. 26, 2002.
78. James Watson, "Fixing the Human Embryo Is the Next Step for Science," *Independent* (London), April 16, 2001.
79. John Campbell and Gregory Stock, *Engineering the Human Germline* (New York: 2000).
80. John Campbell and Gregory Stock, Summary Report, "Engineering the Human Germline" June 1998, p. 15, *www.ess.ucla.edu/huge/report.*
81. Campbell and Stock, "Engineering the Human Germline."
82. Nell Boyce, "And Now, Ethics for Sale? Bioethicists and Big Bucks. Problem City?" *U.S. News & World Report,* July 30, 2001.
83. Sheryl Gay Stolberg, "Bioethicists Find Themselves the Ones Being Scrutinized," *New York Times,* Aug. 2, 2001.
84. Daniel Callahan, "Calling Scientific Ideology to Account," *Society,* May/June 1996, p. 19.
85. William Saletan, "The Ethicist's New Clothes," *www.slate.com,* July 17, 2001.
86. Barbara Katz Rothman, *Genetic Maps and Human Imaginations* (New York: 1998), p. 37.
87. Saletan, "Ethicist's New Clothes."
88. Paulina Borsook, *Cyberselfish* (New York: 2000), pp. 35, 32, 16.
89. Paul and Cox, *Beyond Humanity,* p. 427.
90. Greg Burch, "Progress, Counter Progress, and Counter Counter Progress," speech, Extro 5, June 16, 2000, *www.gregburch.net.*

91. Stock, *Redesigning Humans*, pp. 110–11.
92. Silver, *Remaking Eden*, pp. 8–9.
93. Richard Lemonick, "Designer Babies," *Time*, Jan. 11, 1999.
94. William Saletan, "Fetal Positions," *Mother Jones*, May/June 1998, p. 59.
95. Ed Frauenheim, "Small Changes," *Techweek*, Feb. 7, 2000, *www.techweek.com.*
96. Lester Thurow, *Creating Wealth: The New Rules for Individuals, Companies, and Nations in a Knowledge-Based Economy* (New York: 1999), p. 33.
97. Michael Baumberg and Don Yaeger, "Over the Edge," *Sports Illustrated*, April 14, 1997.
98. Martine Rosenblatt, *Unzipped Genes* (Philadelphia: 1997), p. 66.
99. Campbell and Stock, "Summary Report," p. 14.
100. Carl Pope, "Between Scylla and Charybdis," keynote address, National Abortion and Reproductive Rights Action League, Nov. 9, 2001.
101. Andrew Kimbrell, *The Human Body Shop* (San Francisco: 1993), p. 244.
102. Peter Singer, *A Darwinian Left: Politics, Evolution, and Cooperation* (New Haven, Conn.: 2000), p. 63.
103. Andrew Piper, "Project Ubermensch: German Intellectuals Confront Genetic Engineering," *Lingua Franca*, Jan. 2000, pp. 74–75.
104. Donna Haraway, "The Ironic Dream of a Common Language for Women in an Integrated Circuit," History of Consciousness Board, University of California at Santa Cruz, Oct. 1983.
105. Chris Hables Gray, "Introduction," in Gray, ed., *Cyborg Handbook*, p. 7.
106. Donna Haraway, "A Cyborg Manifesto: Science, Technology, and Socialist-Feminism in the Late Twentieth Century," in Haraway, *Simians, Cyborgs, and Women* (New York: 1991) p. 149ff.
107. Robert Weinberg, "Of Clones and Clowns," *Atlantic Monthly*, June 2002.
108. Engineering the Human Germline Symposium, summary report, July 1998, *www.ess.ucla.edu/huge.*
109. Extropian digest, Nov. 28, 2001.
110. Chris Mooney, "When Left Becomes Right," *American Prospect*, Jan. 29, 2002.
111. Center for Genetics and Society Annual Report, Dec. 31, 2001, *www.genetics-and-society.org.*
112. Nigel Cameron and Lori Andrews, "Cloning and the Debate on Abortion," *Chicago Tribune*, Aug. 8, 2001.
113. Sheryl Gay Stolberg, "Some for Abortion Rights Lean Right in Cloning Fight," *New York Times*, Jan. 24, 2002.
114. Cameron and Andrew, "Cloning."
115. Pope address, NARRAL.
116. E. J. Dionne, Jr. "Unlikely Allies on Cloning," *Washington Post*, Aug. 3, 2001; Center for Genetics and Society Annual Report.

117. Stolberg, "Some for Abortion Rights Lean Right."
118. Chris Mooney, "Senator Sam Brownback, Anti-Corporate Leftist," *American Prospect Online,* Feb. 25, 2002, *www.prospect.org/WebFeatures/2002/02/mooney.*
119. Francis Fukuyama, *Our Posthuman Future* (New York: 2002), pp. 191, 209.
120. Leon Kass, "Preventing a Brave New World," *The New Republic,* May 17, 2001.

CHAPTER FIVE: ENOUGH

1. Gregory Stock, *Redesigning Humans* (Boston: 2002), p. 159; Hughes, "Embracing Change with Four Arms," *www.changesurfer.com,* p. 8; "Biotech Futures," *Wild Duck Review,* Summer 1999.
2. Hughes, "Embracing Change."
3. Ed Regis, "Meet the Extropians," *Wired,* Oct. 1994.
4. "A Letter to Mother Nature," *www.maxmore.com.*
5. Damien Broderick, *The Spike* (New York: 2001), p. 192.
6. "Letter to Mother Nature."
7. Broderick, *The Spike,* p. 199.
8. Nick Bostrom, "A Transhumanist Perspective on Genetic Enhancements," *www.nickbostrom.com,* p. 1.
9. Jeremy Rifkin, *The Biotech Century* (New York: 1998), p. 170.
10. Henri Bergson, *Creative Education* (New York: 1931), chapter 11.
11. Stock, *Redesigning Humans,* p. 171.
12. John Glassie, "Flesh, Robots, and God," *Salon,* Feb. 25, 2002.
13. Andrew Kimbrell, *The Human Body Shop* (San Francisco: 1993), p. 234.
14. Andrew Kimbrell, in Stephanie Mills, ed., *Turning Away from Technology* (San Francisco: 1997), p. 76.
15. Rodney Brooks, *Flesh and Machines* (New York: 2002), p. 167ff.
16. Ibid., p. 160.
17. Gregory S. Paul and Earl Cox, *Beyond Humanity: CyberEvolution and Future Minds* (Rockland, Mass.: 2001), p. 403.
18. Brooks, *Flesh and Machines,* p. 164.
19. Paul and Cox, *Beyond Humanity,* p. 412.
20. Robert Frost, "There Are Roughly Zones," in Edward Connery Lathem, ed., *The Poetry of Robert Frost* (New York: 1979), p. 305.
21. Erazim Kohák, *The Embers and the Stars* (Chicago: 1984), p. 91.
22. Michael Paulson, "One Dam Down, Others in Line," *Seattle Post-Intelligencer,* July 2, 1999.
23. Elizabeth Grossman, *Watershed: The Undamming of America* (New York: 2002), pp. xiii, 21.
24. Lester Thurow, *Zero Sum Society* (New York, 1981) p. 120.
25. Kohák, *Embers and Stars,* p. 103.
26. Matthew 4:4.

27. Anna Foerst, remarks, symposium on "Are We Becoming an Endangered Species?," National Cathedral in Washington, Nov. 19, 2001.
28. Edmund Furse, "The Theology of Robots," *www.smallwonder.hispeed.com/MindsSouls/TheologyRobots.html.*
29. J. Madeline Nash, "Cloning's Kevorkian," *Time,* Jan. 19, 1998.
30. Genesis 2:2.
31. Robert Pack, *Affirming Limits* (Amherst, Mass.: 1985), p. 6.
32. Ibid., p. 14.
33. Ibid., p. 8.
34. Ibid., p. 75.
35. William Shakespeare, *Macbeth,* Act 5, Scene 5, lines 26–28.
36. Leonard Hayflick, *How and Why We Age* (New York: 1994), p. 338.
37. "The Most Important Invention in the Past 2000 Years," *www.edge.org/documents/invention.*
38. Ibid.
39. Ray Kurzweil, *The Age of Spiritual Machines* (New York: 1999), p. 19.
40. Hans P. Moravec, *Robot: Mere Machine to Transcendent Mind* (New York: 1999), p. 4.
41. Marvin Minsky, "Will Robots Inherit the Earth?" *Scientific American,* Oct. 1994.
42. Damien Broderick, *The Last Mortal Generation* (Sydney: 1999), p. 56.
43. Stock, *Redesigning Humans,* p. 23.
44. Robert Ettinger, *The Prospect of Immortality* (Garden City, N.Y.: 1964), p. 98.
45. Stock, *Redesigning Humans,* pp. 92–93.
46. William Shakespeare, *The Tempest,* Act 5, Scene 1, ll. 200–201.
47. Brooks, *Flesh and Machines,* p. 238.
48. Stock, *Redesigning Humans,* p. 198.
49. J. Hughes, "Embracing Change with All Four Arms," *www.changesurfer.com.*
50. Mark Alan Walker, "Prolegomena to Any Future Philosophy," *www.transhumanist.com.*
51. Lee M. Silver, *Remaking Eden* (New York: 1999), pp. 248–50.

Acknowledgments

This book ranges far and wide, obviously, and could not have been written without an equally wide-ranging group of colleagues and friends. I am grateful to David Little and Harvard's Center for the Study of Values and Public Life for the initial year of support, and to the college's Tim Weiskel for the seminar where I tried the initial versions of these ideas on a wise audience. Therese Fitzsimmons, Robert Massie, Dan Smith, Fred Small, Mary Hunt, Jim McCarthy, and James Carroll, among many other Boston colleagues, helped me get started in the right direction.

The crucial research and writing were done with the extremely gracious support and superb facilities of Middlebury College. Nan Jenks Jay, Robert Schine, John McCardell, Ron Leibowitz, Jim Larrabee, and Alison Byerly all helped smooth my way, as did the top-notch library staff, headed by Louise Zipp. My colleagues in Middlebury's dynamic Environmental Studies program include dear old friends like John Elder (who pointed me in very useful directions and provided a crucial early reading), Chris Klyza, Steve

Trombulak, and a host of sharp and engaging new collaborators: Helen Young, Rich Wolfson, John Isham, Don Mitchell, Becky Gould, David Rosenberg, Bob Osborne, Janet Wiseman, Connie Bisson, and others. I've also been lucky enough to meet many wise students willing to talk things over with me, including Andrew Savage, Kaitlin Gregg, Dane Springmeyer, Bennet Konesni, Ben Gore, Jen Marlow, Simon Isaacs, and Caleb Elder. Other necessary Middlebury colleagues include Chris Shaw, Sue Kavanagh, Missy Foote, Dick Nessen, Kathy Nessen, Bobbi Loney, Gerry Loney, Sharon Pinsonneault, Jody Woos, Nancy Cyr, John Barstow, Kate Gridley, Don Stratton, Paul Bortz, and Catherine Nichols. Without Warren and Barry King and Jim and Sheila Hutt, this book would not have been written. Warm thanks, also, to Jeremy Grip.

My home for many years has been in the Adirondack Mountains, and many people there have assisted me in one way or another. Jackie and Nick Avignon, Gary and Kathy Wilson, Amanda Smith-Socaris, Lisa Spilde, Michael Dabroski, Russell Puschak, Kate Gardner, Barb Lemmel, Mitch Hay, Harriet Barlow, Ben Strader, Betsy Folwell, Richard Mayeux, Peter Bauer, Kathleen Collins, Jim and Marcie Sonneborn, Jack and Mary Jean Burke, Jim Gould, Roger Beaudet: Thank you very much. Sam and Lisa Verhovek and Shawn and Michael Considine were, as always, crucial.

Far too little critique of these technologies has yet emerged. Wendell Berry has thought longer and better about human life in a runaway economy and culture than anyone I can name; his influence on these pages is deep. David Abram has also played a critical role, thanks to the insights I gleaned from his writings and those garnered from many wide-ranging conversations in the course of my research. Nancy Green and Stuart Newman also provided exceptionally helpful readings of the text, especially those sections on genetic engineering. For many years, Jeremy Rifkin and Leon Kass have provided the most persistently skeptical critiques of human "enhancement." Casey Walker also dealt with this subject early on. As the text makes clear, I am deeply indebted to Bill Joy for his article outlining the practical dangers, and to Francis Fukuyama's volume *Our Posthuman Future,* which clarifies the political perils. The Center for Genetics and Society, headed by Rich Hayes and Marcy Darnovsky, was one of the first places I turned for help in research, and they graciously opened their files to me. I am also, oddly, very grateful to the proponents of these technologies, who have made no attempt to hide their expectation that the world will change forever if they are allowed to progress unhindered: Gregory Stock, Lee Silver, Ray Kurzweil, Rodney Brooks, Hans Moravec, Marvin Minsky, and many other engineer-writers have forthrightly stated their goals, and I trust they will welcome being forthrightly engaged in return.

My editors at Times Books—David Sobel and John Sterling—and their colleagues, including Heather Rodino, Tracy Locke, Elizabeth Shreve, Maggie Richards, Raquel Jaramillo, Robin Dennis, Jolanta Benal, Kenn Russell, Heather Florence, Marcy Beller, and Chris O'Connell, allowed themselves to be excited by a serious and thorny topic; that makes them rare and delightful publishers. As always, the untoppable Gloria Loomis shepherded this book from start to finish, with the help of her colleague Katherine Fausset.

In part this is a book about the connections between past and future generations. Ray Karras, twenty-five years ago, taught me how to argue. My mother plays a large part in everything I do, as does the memory of my father. Tom and Kristy and Ellie, too. And, of course, Sue and Sophie. They offer me my clearest glimpses of a perfection that has nothing to do with being perfect, and everything to do with being deeply, wonderfully human.

Index

ABOUT THE AUTHOR

Bill McKibben writes regularly for *The New York Review of Books, The New York Times, The Atlantic, Outside, Harper's Magazine,* and many other publications. His first book, *The End of Nature,* was published in 1989 after being excerpted in *The New Yorker;* it was a national bestseller and appeared in twenty foreign editions. His other books include *The Age of Missing Information, Maybe One,* and *Long Distance: A Year of Living Strenuously.* A scholar in residence at Middlebury College, he lives with his wife, the writer Sue Halpern, and their daughter in the mountains above Lake Champlain.